Integrating SAP BusinessObjects BI 4.2 with SAP BW and SAP HANA

insiderBOOKS is a new line of interactive eBooks from the publisher of SAPinsider. With insiderBOOKS you can evolve your learning and gain access to the very latest SAP educational content.

The team at insiderBOOKS has created a brand-new solution for the digital age of learning. The insiderBOOKS platform offers affordable (and, in many cases, free!) eBooks that are updated and released chapter by chapter to ensure that you're always reading the freshest content about the latest SAP technologies.

Each eBook provides more than what you can get from the typical textbook—including high-res images, cloud-based material (access your content anywhere!), videos, media, and much more. Plus, thanks to our generous sponsors, you get the opportunity to qualify for free eBooks.

insiderBOOKS is Reading Redefined.

Please visit our website for additional information:
www.insider-books.com

Ingo Hilgefort

Integrating SAP BusinessObjects BI 4.2 with SAP BW and SAP HANA

From the publisher of SAPinsider

Integrating SAP BusinessObjects BI 4.2 with SAP BW and SAP HANA
by Ingo Hilgefort

First edition: August 2016.
ISBN 978-0-9974291-1-4

Published by Wellesley Information Services, LLC (WIS), 20 Carematrix Drive, Dedham, MA, USA 02026.

Publisher	Melanie A. Obeid
Product Director	Jon Kent
Acquisitions Editor	Jawahara Saidullah
Editor	Andrea Haynes
Copyeditors	Margaret Hein and Matt Walsh
Art Director, Cover Designer	Jill Myers
Production Artist	Kelly Eary
Production Director	Randi Swartz
Marketing	Michelle Burke

About the Author

 **Ingo Hilgefort** started his career in 1999 with Seagate Software/Crystal Decisions as a trainer and consultant. He moved to Walldorf for Crystal Decisions at the end of 2000 and worked with the SAP NetWeaver BW development team integrating Crystal Reports with SAP NetWeaver BW. He then relocated to Vancouver in 2004 and worked as a product manager/program manager (in engineering) on the integration of BusinessObjects products with SAP products.

Ingo's focus is now on the integration of the SAP BusinessObjects BI suite with SAP landscapes, such as SAP BW and SAP BW on SAP HANA, focusing on end-to-end integration scenarios. In addition to his experience as a product manager and in his engineering roles, Ingo has been involved in architecting and delivering deployments of SAP BusinessObjects software in combination with SAP software for a number of global customers, and has been recognized by the SAP Community as an SAP Mentor for SAP BusinessObjects- and SAP integration-related topics.

Currently, Ingo is the vice president of product management and product strategy at Visual BI Solutions, working on extensions to SAP's product offerings such as SAP BusinessObjects Design Studio and SAP BusinessObjects Lumira. You can follow him on Twitter at @ihilgefort.

ACKNOWLEDGEMENTS

Special thanks go to the team from insiderBOOKS, who made it possible for me to focus on the writing and not to worry about style, layout, or publishing a book. My thanks in particular to Andrea Haynes, Margaret Hein, Melanie Obeid, and Jon Kent, who helped make this happen and for making the review process as smooth as possible. Also, I can't forget Tammy Powlas for being a great reviewer of the material and Gopal Krishnamurthy, CEO of Visual BI Solutions, for providing the environment for me to focus on the writing.

Many thanks also to Gaby, Ronja, Sally, and Zoe for the well-needed writing breaks.

Contents in Brief

Contents in Full

Chapter 3   SAP BusinessObjects Analysis, edition for Microsoft Office (Analysis Office) 131

Chapter 4   SAP BusinessObjects Design Studio (Design Studio) 163

About This Book

The integration of SAP systems and SAP BusinessObjects BI tools has constantly improved since 2008, when SAP acquired BusinessObjects. The focus of the overall SAP BusinessObjects BI portfolio has shifted from Crystal Reports and Web Intelligence in the early years, to Xcelsius (now called SAP Dashboards) and SAP BusinessObjects Explorer, and, in the most recent discussions, to SAP BusinessObjects Analysis, edition for Microsoft Office (Analysis Office), SAP BusinessObjects Lumira, and SAP BusinessObjects Design Studio.

Even though the integration of SAP BusinessObjects BI with SAP systems—such as SAP BW and SAP HANA—has been around for several years, companies are still seeking a good overview on the integration itself. They want to know how to best leverage the different SAP BusinessObjects BI tools and the existing assets from their SAP landscape. That is what this book is trying to accomplish by giving you a simple and practical overview of the most important aspects of the integration of SAP BusinessObjects BI with SAP.

I hope this book gives you a simple but sufficiently technically detailed overview of what you can do today with the latest SAP BusinessObjects BI 4.2 software in combination with your SAP landscape.

Target Group

The book is written for those looking for easy-to-follow instructions on how to use and deploy the SAP BusinessObjects BI software in combination with an SAP landscape. The book focuses on putting you in a position to leverage an SAP BusinessObjects BI 4.2 system on top of your SAP system, to install and configure the software, and to create your first reports with tools such as SAP BusinessObjects Web Intelligence, SAP BusinessObjects Analysis, edition for Microsoft Office (Analysis Office),

SAP Design Studio, and SAP Lumira software. It is not the goal of this book to make you an SAP BusinessObjects expert or to explain every detailed aspect of the SAP BusinessObjects software, because several other resources already fulfill such a need.

As a reader of this book, you should have some previous knowledge of SAP BW and SAP ERP. On the SAP BusinessObjects side, I tried to keep the need for previous knowledge as minimal as possible. You should be able to follow this book even without any prior SAP BusinessObjects knowledge, but you should consider further product documentation and training.

Focus Areas

Note that this book focuses on those products that are currently strategically important for SAP as part of the overall SAP BusinessObjects BI portfolio:

- Analysis Office
- SAP BusinessObjects Design Studio
- SAP Lumira
- SAP BusinessObjects Web Intelligence

Technical Prerequisites

All steps and examples in this book are based on the SAP BusinessObjects BI 4.2 release in combination with an SAP BW 7.4 system.

You can download the SAP BusinessObjects software from the SAP Service Marketplace. The book includes practical step-by-step instructions, so I highly recommend that you download the following components so you can follow all outlined steps:

- SAP BusinessObjects BI Platform 4.2
- Analysis Office 2.3
- SAP BusinessObjects Design Studio 1.6 Support Package 02
- SAP Lumira 1.31

You should ensure that you have access to an SAP BW and an SAP HANA system so that you can follow the examples.

Structure of the Book

Here is an overview of the chapters' contents.

Chapter 1 — SAP BusinessObjects BI 4.2 with SAP BW and SAP HANA

Chapter 1 introduces you to the 4.2 release of the SAP BusinessObjects BI platform and the suite of BI client tools that you will use in the following chapters. You'll get a brief overview of the main parts of the SAP BusinessObjects BI platform and learn which of the SAP BusinessObjects BI tools is best suited for which kind of user audience.

Chapter 2 — Installation and Configuration

In Chapter 2 you'll learn how to install and configure the 4.2 release of the SAP BusinessObjects BI platform and client components. You'll receive step-by-step instructions on the installation of the software, the configuration steps on the SAP BusinessObjects BI platform side, and the installation of the SAP BusinessObjects BI client tools.

Chapters 3 to 6 — SAP BusinessObjects BI Clients

In Chapters 3 to 6 you'll learn more about the capabilities of Analysis Office, SAP Design Studio, SAP Lumira, and SAP BusinessObjects Web Intelligence. You'll receive an overview on how each product is able to connect to your SAP systems and how it is able to support the different metadata available. In addition, each of these chapters includes step-by-step instructions on how to create your first reports or dashboards with the product.

Chapter 7 — Integration Roadmap

Chapter 7 provides a brief overview of the upcoming integration topics for those products discussed in this book.

CHAPTER 1

Introduction to SAP BusinessObjects BI 4.2

In this chapter we will provide a quick overview of the SAP BusinessObjects BI platform, as well as learn more about the different SAP BusinessObjects BI client products. After the overview, we will take a look at the business requirements on which each of the BI client products is focused, as well as the types of user audiences.

1.1 Introduction to the SAP BusinessObjects BI Platform

In this section, we will take a quick look at the SAP BusinessObjects BI platform and the SAP BusinessObjects client tools that we discuss throughout the book. The purpose of this section is not to give an in-depth overview of the SAP BusinessObjects architecture; the intention is to provide an overview of the architecture with enough details so that you can install, deploy, and configure the software. Figure 1.1 shows the different tiers of the SAP BusinessObjects BI platform.

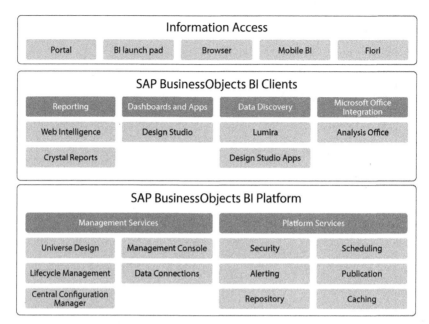

Figure 1.1 SAP BusinessObjects BI platform

1.1.1 Information Access

For Information Access, the SAP BusinessObjects BI platform includes a large set of tools and options. The most common user interface (UI) is the BI launch pad. The BI launch pad provides the user with a complete set of capabilities to leverage all the features and functions of the BI client tools, and delivers functionality such as viewing, scheduling, and broadcasting of reports and analytics to the end user. In addition, the SAP BusinessObjects BI platform delivers an out-of-the-box application, so that content can be consumed on a mobile device, called SAP Mobile BI.

Part of the user interaction layer is also the integration into the different portal environments, such as an integration into SAP Enterprise Portal or Microsoft SharePoint. Access to your BI-related content on the SAP BusinessObjects BI platform can also be integrated with the SAP Fiori launchpad.

1.1.2   SAP BusinessObjects BI Client Tools

As part of the SAP BusinessObjects BI suite, SAP is delivering a set of BI client tools that can be categorized into four categories:

- Reporting
- Dashboards and apps
- Data discovery
- Microsoft Office integration

In each of these categories, SAP is delivering a set of products focused on this particular functionality. In the reporting category you can leverage SAP Crystal Reports and SAP BusinessObjects Web Intelligence to provide your users with the typical reporting functionality. For example, you can create predefined reports using Crystal Reports or leverage features such as printing and exporting using Crystal Reports or Web Intelligence.

In the dashboards and apps category, the strategic tool from SAP is SAP BusinessObjects Design Studio, replacing SAP Dashboards (formerly Xcelsius) and the SAP Web Application Designer.

In the area of data discovery, SAP is positioning SAP BusinessObjects Lumira as well as applications created using SAP BusinessObjects Design Studio. In the area of integration with Microsoft Office you can leverage SAP BusinessObjects Analysis, edition for Microsoft Office (Analysis Office).

We will go into further details on each of the BI client tools in a later section in this chapter, as well as in the individual chapters for each of the BI tools.

1.1.3   SAP BusinessObjects BI Platform Services

The SAP BusinessObjects BI platform is delivering a set of management tools and platform services to you, providing you with capabilities such as scheduling a report and establishing data connectivity to your data warehouse.

1.1.3.1   Management Tools

In the area of the management tools, the BI platform delivers several options to manage the system itself and manage the integration with other system landscapes.

Central Management Console (CMC)

The CMC is a web-based tool that allows you to administer and configure your SAP BusinessObjects BI platform. The following list represents some of the main tasks you can perform using the CMC:

- Create, configure, and manage users and user groups.
- Create, configure, and manage services of your platform.
- Integrate with other user authentication providers, such as Lightweight Directory Access Protocol (LDAP), Microsoft Active Directory, and SAP Authentication.
- Assign object security to users and user groups.
- Set up and configure scheduling and publications.
- Administer and manage BI content and content categories.

Central Configuration Manager (CCM)

The CCM provides the administrative functionality to configure and manage the services of your SAP BusinessObjects BI platform system. You can use the CCM to start, stop, enable, or disable services and to perform configuration steps on those services. In contrast to the CMC, which is available as a web client, the CCM is available as a Windows-only client.

Lifecycle Management Tool

The SAP BusinessObjects BI platform delivers a Lifecycle management console that allows you to move content and dependent objects from your development environment to your test and quality environment and finally to your production environment. You can manage different versions, data connectivity, content dependencies, and the promotion of content objects between different systems.

Data Connections and Universe Design

Part of the SAP BusinessObjects BI management services for the administrator is the ability also to set up data connections to a large variety of source systems, such as SAP HANA, SAP BW, and SAP S/4HANA as well as non-SAP data sources such as Microsoft SQL, Oracle DB, and IBM DB2—just to mention a few. In addition, the BI platform also provides you with the Information Design Tool, which allows you to establish Universes—a business semantic layer—based on those data sources.

1.1.3.2 *Platform Services*

Platform services deliver core BI functionality to the end user and system administrator. The following is a brief explanation of the main components of the platform services.

Central Management Server (CMS)

The CMS is the "heart" of the system; it manages and controls all the other services. In addition, the CMS manages access to the system database. The system database contains all the information about users, groups, server configurations, available content, available services, and so on. The main tasks of the CMS are:

- Maintaining security by managing users and user groups and the associated groups configured in the SAP BusinessObjects platform.

- Managing objects by keeping track of all objects hosted in the platform, and managing the physical location of those objects and the object definition by using the system database.

- Managing services by constantly validating the status of each service and the overall list of available services. In addition, the CMS is able to handle load balancing to allow for enhanced scalability and better use of hardware.

- Auditing by keeping track of any event inside the platform and thus allowing the administrator to base further deployment considerations on actual information from the usage of the system.

File Repository Services

Each deployment of the SAP BusinessObjects platform has an input and an output file repository service. The input service is responsible for storing the content available in the platform, except instances (scheduled reports), which are managed by the output file repository service.

Processing Tier

The processing tier of the SAP BusinessObjects platform includes (along with other components) the following main components:

Job Processing Services

Job processing services are responsible for fulfilling requests from the CMS to execute a prescheduled job for a specific report. To be able to run the job successfully, these services require access to the underlying data sources.

Cache Services

Cache services are used in combination with the job processing services. When a request from a user can be fulfilled with cached information, there is no need to use the job processing service. Otherwise, the cache services hand over the request to the job processing services.

Viewing Processing Services

Viewing processing services are responsible for retrieving the content objects from the file repository services, executing the content against the data source, and showing the actual content to the end user via the different viewer types.

Server Intelligence Agent (SIA)

The SIA allows you to simplify some administrative tasks, such as adding or removing services or starting and stopping services. For example, you could assign all the processing services to a specific SIA, which then allows you to start and stop all those services with a single command.

Adaptive Processing Server (APS)

The APS is a generic server that is able to host a set of services that is responsible for processing requests, such as processing an SAP BusinessObjects Design Studio request. By default, your SAP BusinessObjects BI platform is installed with one APS per host system in a distributed landscape. It is a best practice to configure additional APSs to meet your business requirements for performance and scalability.

In this section we provided an overview on the overall SAP BusinessObjects BI platform. In the next section we will review the strategic direction for the SAP BusinessObjects BI client tools, as well as learn more about each of the tools.

1.2 Introduction to SAP BusinessObjects BI Clients

In this section we will briefly cover those SAP BusinessObjects BI tools that we will discuss in more detail in the following chapters in this book. The main purpose of this section is to give you a brief introduction to each of the tools and to explain their main uses.

1.2.1 SAP BusinessObjects Web Intelligence

SAP BusinessObjects Web Intelligence is a BI tool focusing on the concept of self-service reporting, and providing the end user with the ability to create ad-hoc new reports or to change existing reports based on new business requirements. SAP BusinessObjects Web Intelligence empowers the end user to answer business questions using a very simple and intuitive UI and typically providing access to a broader range of data. SAP BusinessObjects Web Intelligence allows the end user to dynamically create data-relevant queries; apply filters to the data; sort, slice, and dice through data; drill down; find exceptions; and create calculations.

Using SAP BusinessObjects Web Intelligence, you can easily create a simple sales report showing revenue broken down according to several dimensions, as shown in Figure 1.2.

Sales Report

Year	Quarter	State	City	Sales revenue	Quantity sold
2004	Q1	California	Los Angeles	$308,928	2,094
2004	Q1	California	San Francisco	$210,292	1,415
2004	Q1	Colorado	Colorado Springs	$131,797	921
2004	Q1	DC	Washington	$208,324	1,467
2004	Q1	Florida	Miami	$137,530	924
2004	Q1	Illinois	Chicago	$256,454	1,711
2004	Q1	Massachusetts	Boston	$92,596	609
2004	Q1	New York	New York	$555,983	3,717

Figure 1.2 Simple SAP BusinessObjects Web Intelligence report

More importantly, with a few clicks you'll be able to change the report to a sales report showing the top 10 states based on revenue with a chart showing the top 10 cities based on revenue (see Figure 1.3).

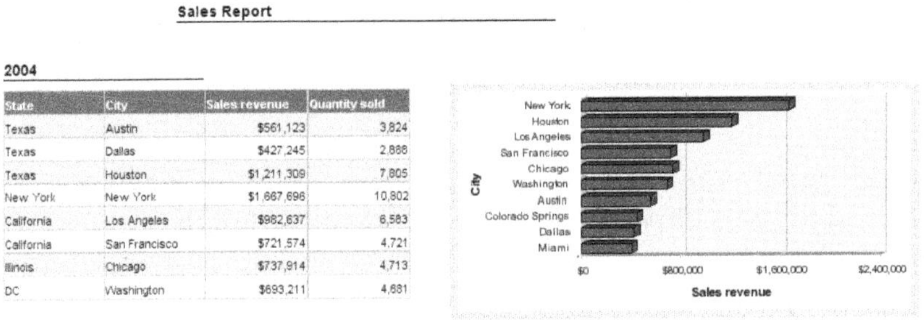

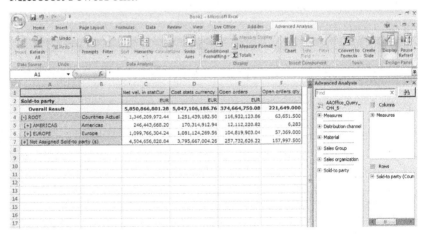

Figure 1.3 SAP BusinessObjects Web Intelligence report with chart

SAP BusinessObjects Web Intelligence allows you to reduce your IT department's workload of creating or changing reports and to provide your end users with a real self-service reporting environment.

1.2.2 SAP BusinessObjects Analysis, edition for Microsoft Office (Analysis Office)

Analysis Office is an effective addition to the overall BI client portfolio as part of the SAP BusinessObjects 4.2 Suite. Analysis Office (see Figure 1.4) is the premium successor to the SAP Business Explorer (BEx) Analyzer, and allows you to use analytical capabilities inside Microsoft Excel and Microsoft PowerPoint.

Figure 1.4 Analysis Office

In addition to being the premium successor for the SAP BEx Analyzer product, Analysis Office is also the key product providing you, on the one hand, with the BI workflows inside the Microsoft Office environment, and on the other hand, it is becoming the successor for SAP BusinessObjects Live Office as well as the SAP Enterprise Performance Management (EPM) add-in.

1.2.3 SAP BusinessObjects Design Studio

SAP BusinessObjects Design Studio is the premium successor for the SAP Web Application Designer (WAD), and it is also the premium successor for SAP BusinessObjects Dashboards. SAP BusinessObjects Design Studio is a developer-focused environment that allows the designer to create professionally authored dashboards and applications (see Figure 1.5).

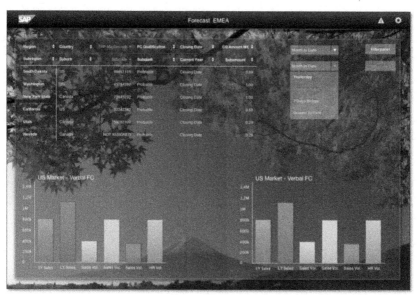

Figure 1.5 SAP BusinessObjects Design Studio sample

SAP BusinessObjects Design Studio renders the output of the dashboard and application in HTML5. Therefore, the content created by SAP BusinessObjects Design Studio is available for the desktop, for a browser, and for a mobile device.

CHAPTER 1

1.2.4 SAP Lumira

SAP Lumira focuses on the business user as the audience type and provides an environment that allows the user to acquire the data from the source, perform a set of data manipulations (such as grouping data together), visualize the information (see Figure 1.6), and share the findings with a larger audience.

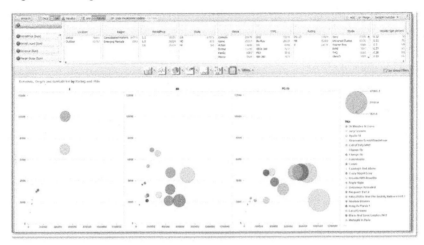

Figure 1.6 SAP Lumira example

SAP Lumira focuses on the data discovery and visualization as part of the overall BI portfolio and gives your business user the ability to perform self-service-focused analysis.

1.2.5   Semantic Layer

The Information Design Tool allows you to create connections to your data sources and to create Universes that you can then provide to business users to analyze data in a more user-friendly way. A universe provides an easy-to-understand and non-technical interface for your users so they can focus on analyzing and sharing the data by using common business terms.

The Information Design Tool provides you with a graphical interface allowing you to define connections to relational database management systems (RDBMS) and multidimensional data structures (online analytical processing [OLAP]), and to define the business semantics on top of the data source.

In this section we provided a short overview on those BI client tools from the SAP BusinessObjects BI portfolio that we will discuss in more detail later in this book. In the next section, we will learn more about how each of the BI client tools will help us to fulfill requirements from different types of users.

1.3 Strategic Direction for the SAP BusinessObjects BI Portfolio

In this section, we will take a look at SAP's strategic direction of the SAP BusinessObjects BI portfolio. We also will learn more about how we can match the different BI tools against a set of capabilities and how we can map the different BI tools to different types of audiences.

1.3.1 SAP BusinessObjects BI Portfolio Simplification

Before we start looking at how we can map the SAP BusinessObjects BI tools against some key capabilities and different types of consumers in your organization, it is important to first understand the strategic direction from SAP when it comes to the overall SAP BusinessObjects BI portfolio and how SAP positions the different BI clients against each other.

Figure 1.7 shows how SAP is planning to simplify the overall SAP BusinessObjects BI portfolio over the next couple of months.

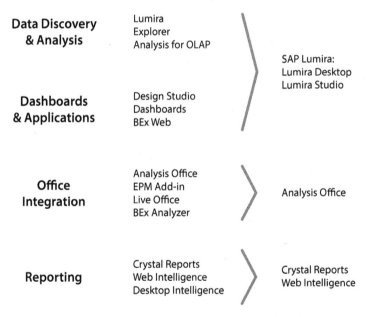

Figure 1.7 SAP BusinessObjects BI portfolio simplification

Note how SAP is consolidating the overall list of BI clients in the SAP BusinessObjects BI portfolio. As you will notice, only a small set of BI clients is shown moving forward in SAP's strategic direction. The products shown in the middle part of Figure 1.7 are those that have been available for some time and some of them—for example, Desktop Intelligence—are already treated as legacy products. The products shown on the right side of Figure 1.7 are those that SAP sees as the strategic direction moving forward. You will notice that SAP BusinessObjects Design Studio is not shown on the right side of the image, but also notice that SAP Lumira Studio is listed. Starting with the next major release, planned for Q1 2017, SAP Lumira and SAP BusinessObjects Design Studio will both be branded under the SAP Lumira product family, so SAP BusinessObjects Design Studio will become SAP Lumira Studio.

Figure 1.8 provides a slightly different view. It shows the products assigned to the different categories—Data Discovery & Analysis, Dashboards & Applications, and Reporting—as well as the differentiation based on whether the product is more of a self-service functionality or one that requires a specific person or team to create the report or dashboard before it is shared with the business user audience.

Professionally Authored		
	Design Studio	Crystal Reports
Data Discovery & Analysis	Dashboards & Applications	Reporting
Self-Service		
SAP Lumira Analysis Office Design Studio Applications	Design Studio Applications SAP Lumira	Web Intelligence

Figure 1.8 SAP BusinessObjects BI tool positioning

As you can see, the SAP Lumira and Design Studio applications are positioned in the Data Discovery area as well as in the Self-Service area, which should not come as a surprise as both products play a key role in these important areas.

1.4 Mapping SAP BusinessObjects BI Tools to Capabilities

After looking at the strategic direction from SAP and how the BI clients are positioned in the different categories, we will now take a look at some of the main capabilities of the SAP BusinessObjects BI products and compare them with each other.

In Figure 1.9, you can see a ranking of the SAP BusinessObjects BI tools mapped to several key capabilities.

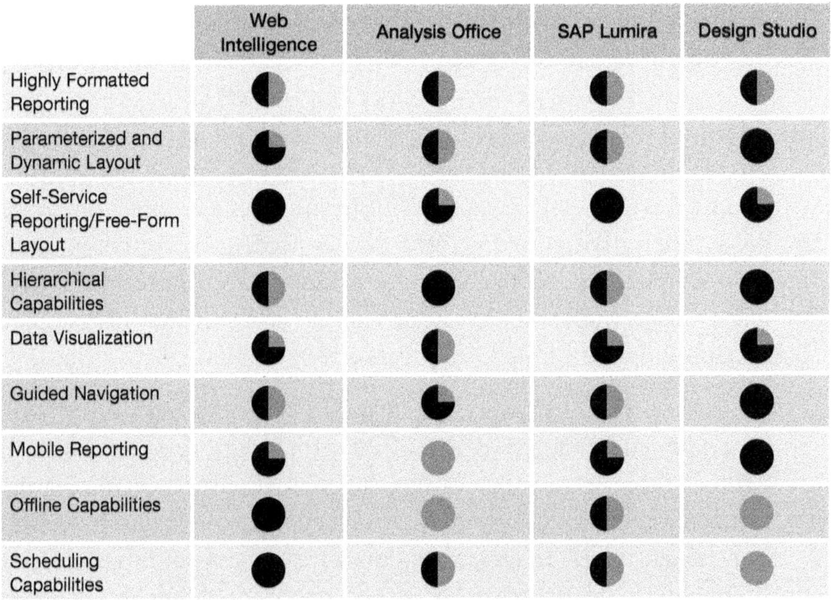

	Web Intelligence	Analysis Office	SAP Lumira	Design Studio
Highly Formatted Reporting	◐	◐	◐	◐
Parameterized and Dynamic Layout	◕	◐	◐	●
Self-Service Reporting/Free-Form Layout	●	◕	●	◕
Hierarchical Capabilities	◐	●	◐	●
Data Visualization	◕	◐	◕	◕
Guided Navigation	◐	◕	◐	●
Mobile Reporting	◕	●	◕	●
Offline Capabilities	●	○	◐	○
Scheduling Capabilities	●	◐	◐	○

Figure 1.9 SAP BusinessObjects BI tools' capabilities

Now let's take a look at the details of these capabilities and provide more background and insight into the ranking.

- Highly Formatted Reporting: In this category, it's important that the tool provides full control over the layout and that you are able to create a report that is going to look identical in all web clients or when it is exported to an external format, such as a portable document format (PDF). The extreme example for this is creating reports that are identical to legal forms. However, formatted layouts can be very useful in other areas as well when it is important to have a well-structured and formatted report, such as a delivery notice, customer invoice, or a balance sheet.

- Parameterized and Dynamic Layout: With parameterized layouts, we're referring to the concept that the layout can be influenced by the consumer of the report simply by changing some parameters. A good example is a report that allows you to see the data grouped, or as a simple list, or as a chart. The user is able to influence the layout of the report by simply setting a value for a parameter that selects one of those options. The other example of a parameterized layout is one that has the capability to show different types of data visualization based on user input—for example, showing a weekly, quarterly, or monthly comparison after the user has selected one of the three options. In addition, part of a parameterized layout is its capability to influence the layout based on defined conditions and the data being retrieved. The simplest example of this functionality is the ability to highlight a key figure based on a value and thresholds. A more complete example is to completely suppress a Top 5 chart for which only three values exist, and thus all would be shown.

- Self-Service Reporting/Free-Form Layout: Self-service reporting (sometimes referred to as free-form layout-driven reporting) allows the user to create or change content without involving the information technology (IT) department to create a new report or make changes to an existing report. The concept of self-service reporting is more of the actual tool functionality than it is the type of reports or analytics that can be created. Self-service reporting is focused on offering the consumer a tool that provides an easy-to-use environment that puts the user in the "driver's seat" of the report—enabling the user to create or edit the report as needed.

- Hierarchical Capabilities: In this category, the tools are compared based on the capability of leveraging an existing hierarchy from SAP BW or SAP HANA, and being able to present the hierarchy properly as part of the report. The tool should not only be able to actually identify the hierarchy, but also to create a hierarchical-organized report; allow formatting of the report based on hierarchical information, such as the hierarchy level; and also recognize things such as a hierarchy variable and hierarchy node variable. In addition, this category also includes the actual hierarchy navigation a consumer of the report can perform.

- Data Visualization: This category focuses on the set of capabilities needed to visualize actual data and to provide interactive capabilities. It is important to recognize that this is not a comparison of all the different charting options of the tools. Charting is one element of the data-visualization capabilities. Other elements include interactive navigation and the ease-of-use of the visualization.

- Guided Navigation: The term "guided navigation" is used to describe the capability to provide ad-hoc analysis and to limit the scope of change for the user, so that the user is only able to change specific parts of the analysis workflow. In addition, guided navigation refers to the functionality that the designer of the analysis workflow can use to create a pre-determined workflow for the actual consumer of the information. Think about a sales management analysis. The user is able to see his Top 10 customers and the Top 10 opportunities in his pipeline on the initial view of the analysis. In addition, he can see the Top 10 opportunities with the highest risk factor of not being closed in the current quarter. Instead of having to navigate through the data, the sales manager can click a button and be "guided" to the second page of his analysis, where he sees more details regarding the Top 10 opportunities that are at risk. You can see that guided navigation helps create a predefined workflow for the consumer that is geared towards anticipating and providing answers to the most commonly asked questions.

- Mobile Reporting: Mobile reporting refers to the opportunity to leverage the content created by the BI tool on a mobile device. Explicitly, this category refers to supporting the mobile device as a native application and not by opening the BI content in a browser session on the mobile device. You will notice that there are different rankings between the tools, which are based on the fact that the support of the different mobile devices, such as Android devices or iPads or BlackBerry, varies between the BI tools.

- Offline Capabilities: Offline in this context refers to the capability to create a report that contains the actual data and still allows the user to navigate in the data, even when the connection to the source system has been disconnected. For example, think of a sales representative who reaches the office on a Monday morning, receives the latest reports, and is still able to navigate in the data while being on the road without a connection to the source system for the rest of the week.

- Scheduling Capabilities: Scheduling capabilities refers to the option to schedule the report or dashboard and to offload the generation of the report or dashboard to a server environment and to a specified timeframe. In addition, scheduling also includes the ability to present the result in different formats and towards different locations.

In Figure 1.10 you can see a comparison of the BI products when it comes to the hierarchical-reporting capabilities. The reason for looking at the hierarchical-reporting capabilities is simply based on the main data sources that we will discuss—SAP BW and SAP HANA—as those frequently leverage hierarchical data structures.

	Web Intelligence	Analysis Office	Design Studio	SAP Lumira
Time-Dependent Hierarchies	✓	✓	✓	✗
Hierarchical Member Selection	✓	✓	✓	✓
Hierarchical Level-Based Selection	✓	✓	✓	✗
Selecting Specific Hierarchy Levels	✓	✓	✓	✗
Skipping Hierarchical Levels	✗	✓	✗	✗
Showing Leaf Members Only	✗	✓	✗	✗
Ranking Hierarchical Data	✗	✓	✓	✗
Exchanging Hierarchies on the fly	✗	✓	✓	✗
Placing Subtotals (above/below)	✗	✓	✓	✗
Expand to Level	✗	✓	✓	✗
Hierarchical Charting	✓	✓	✓	✓

Figure 1.10 Hierarchical-reporting capabilities

You will notice that the two products with the strongest support for the hierarchical-reporting capabilities are Analysis Office and Design Studio. Below you will find a brief description of the capabilities compared in Figure 1.10:

- **Time-Dependent Hierarchies:** This item refers to the support of time-dependent hierarchies or time-dependent hierarchical structures.

- **Hierarchical Member Selection:** Here the user should be able to choose hierarchical nodes or leafs from the actual hierarchical structure and filter the data in that way.

- **Hierarchical Level-Based Selection:** In this case the user should be able to choose the levels that should be included in the reports. So instead of choosing a fixed set of members from the hierarchy, the user chooses the levels.

- **Selecting Specific Hierarchy Levels:** In this scenario the user should be able to receive the list of available hierarchical levels and then choose those levels that will be included in the report itself. The user should also be able to limit the number of levels included in the report.

- **Skipping Hierarchical Levels:** In addition to selecting hierarchical levels, the user should also be able to skip levels as part of the selection process. For example, when given a hierarchy with a maximum of 10 levels the user should be able to select levels 3, 5, and 8 without having to choose all the levels in between.

- **Showing Leaf Members Only:** A hierarchy can be balanced or unbalanced and, in certain scenarios, the user might want to look at the lowest level of the hierarchy itself, displaying only the leaf members. Here it is important that the tool understands the concept of a balanced versus an unbalanced hierarchy.

- **Ranking Hierarchical Data:** In case your data is organized in a hierarchical way, then ranking the data means that the ranking needs to be done based on the hierarchical structure. Ranking data according to the hierarchy means to rank data inside each node against each other, and to rank the nodes on each level against each other.

- Exchanging Hierarchies on the Fly: Often it is crucial for the work-flow for the user to have the ability to change between different types of hierarchies without having to change the actual report or change the underlying data source of the report. Changing to a different hierarchy should be possible as part of the workflow performed by the consumer of the report or dashboard.

- Placing of Subtotals: In the case of a hierarchical report, it is important to be able to decide if the totals and subtotals should be displayed above or below the hierarchical details. A typical example would be a balance sheet displaying the totals and subtotals below the elements of a hierarchy.

- Expand to Level: The Expand to Level is an important time-saving feature as it allows the user to simply expand the complete hierarchy to a specific level, so that the user does not have to open up each hierarchical element level by level.

- Hierarchical Charting: When it comes to hierarchical charting, it is important that the tool understands the hierarchical nature of the data and is able to leverage the hierarchical nature as part of the charting as well. For example, the user should have the ability to drill down from one hierarchical level to the next hierarchical level as part of a drill-down workflow in the chart.

Based on the preceding reporting and analysis categories and the brief descriptions of the compared functionality in each of them, you should now have a much better understanding of the strengths and weaknesses of each tool—even though this section did not show and compare every detail of every tool. The material up to this point is meant to provide you with an overview. In the following chapters, you will see how the tools differentiate from each other, and where you are going to use the products and create the reports, analytics, and dashboards.

In the following section, we'll look at the different user types of BI solutions and how the tools align with those user types. We'll also look at some of the skills that define each user type. It's important to understand both the requirements and the audience when making a decision on actual product usage. For example, you might need a report that provides information along several dimensions, which can be created with SAP BusinessObjects Web Intelligence, SAP Lumira, and Analysis Office.

However, because your audience is a group of not-very-IT-oriented information consumers, you may decide to use prepared reports with a small set of parameters (category: parameterized layout) to offer such functionality. Based on your understanding of report requirements and user type, you decide to go with Web Intelligence.

1.5 Mapping SAP BusinessObjects BI Tools to Personas

It's very important to understand the different user types for the SAP BusinessObjects BI tools and how those user types map to the different products. Before we begin, it should be stated that not every product from the SAP BusinessObjects BI portfolio has been created for each user type. Each tool delivers a specific reporting and analysis user experience to a defined group of user types and has not been created with each user type in mind.

Before we start mapping the BI toolset to the user types, we need to clarify what those user types are and, more importantly, the needs and skills associated with the user types. We must look at this issue from two sides: what the user wants and what he actually needs to do in his day-to-day job. Beyond these two points, you must also consider the skill level of the user. Sometimes the choice of tool can be based solely on product features and functionality, but other times you also have to consider the skills of the person using the tool.

To keep it relatively simple we will break down our user types into four categories:

- Information Consumer/Business User
- Business Analyst/Power User
- Middle Management/Line-of-Business Management
- Decision Maker/"C-Level" Management

You may notice that these user types do not include a role called "Report Designer" or "IT Administrator." The reason is that we want to focus on the consumption of information and how a user can leverage the BI tools to make informed decisions based on the provided information. The person creating the reports and analytics may have a different skillset compared to these user types. We'll show the actual creation of the content in this book, but it is important to understand the consumer types of the reporting and analysis content. By doing so, you'll be better equipped to provide them with the right information in the right tool.

Let's define the typical characteristics and skills of our user types. We'll characterize each user type based on the following:

- What are some typical goals of users working in a BI environment?

- What are some typical tasks for the user type?

- What other software does the user work with on a regular basis?

These tasks and goals are not meant to be specific to an area such as sales or finance, but rather should be seen as generic descriptions of a certain type of task or goal.

User Type: Information Consumer/Business User

- Goals
  - Review regular sales reports and monitor individual accounts and sales status.
  - Review regular account statements to control customer invoices and vendor accounts.
  - Review actual operational measures against goals.
  - Fulfill management requests for information as simply as possible.

- Tasks
  - Find a prepared report, view the information, and print or export the information.
  - Receive and review alerts from prepared reports and analytics.
  - Schedule prebuilt reports and review the resulting information.
  - Use predefined navigation steps and alerts to receive needed information.
  - If required, provide information to the IT department for additional reports and analytics based on the needed information.

- Regularly used software
  - Microsoft Excel
  - Microsoft PowerPoint
  - Microsoft Word
  - Microsoft Outlook
  - Internet browser

User Type: Business Analyst/Power User

- Goals
  - Analyze key performance indicators (KPIs) to find areas for improvement.
  - Create deeper analysis to find details on anomalies.
  - Leverage actual data and historical data to create detailed planning scenarios to enable more realistic forecasting and planning of future company key goals.
  - Leverage the data and tools to provide answers ad hoc to the management and leadership teams so that decisions are based on solid information.

- Tasks
  - Review prepared reports for KPIs and analyze the prepared data for anomalies.
  - Edit existing reports and, if required, create new reports and analytics on the fly to answer related business questions.
  - Share analysis and results with a larger audience and the management/leadership team.
  - Act as the go-to person for the management/leadership teams by providing required analysis for informed decisions.

- Regularly used software
  - Microsoft Excel
  - Microsoft PowerPoint
  - Microsoft Word
  - Microsoft Outlook
  - Microsoft Access
  - Internet browser

User Type: Middle Management/Line-of-Business Management

- Goals
  - Review and analyze regional/departmental goals and KPIs and share findings with upper management.
  - Analyze information to measure the progress towards the set goals and KPIs.
  - Fulfill executive-level requests for information as simply as possible.

- Tasks
  - Jointly set goals and targets on a department/line-of-business level and continuously monitor and review those goals and targets.
  - Regularly analyze operational reports and prepare analytical summaries for upper management.
  - Leverage prebuilt reports to further analyze the detailed information.
  - If needed, leverage the tools to further analyze the information and to measure the current against the agreed goals.
- Regularly used software
  - Microsoft Excel
  - Microsoft PowerPoint
  - Microsoft Word
  - Microsoft Outlook
  - Internet browser

User Type: Decision Maker / "C- Level" Management

- Goals
  - Analyze overall company-wide operational metrics and ensure agreed targets are met.
  - Oversee cross-department/line-of-business performance and evaluate different scenarios for planning and forecasting purposes.
  - Leverage the information for analyzing, monitoring, and planning purposes to continuously improve company performance.
  - Combine the analytics with company strategies and goals, and integrate these strategies and goals into each employee's workflow and goals.
- Tasks
  - Review company-wide metrics (including past, actual, and forecasted values) to make informed decisions and take necessary actions.
  - Set goals and targets for middle management and link them back to company-wide goals and metrics. Continuously monitor and review those goals and targets.

  – Regularly review operational KPIs and look for opportunities to improve operations and profit.

- Regularly used software
  – Microsoft Excel
  – Microsoft Outlook
  – Internet browser

Now that we've defined our user types, we need to map these user types (based on their needs and skills) to the SAP BusinessObjects BI tools.

Figure 1.11 shows the four user types and the optional tools to address their needs. This does not mean, for example, that you cannot use SAP Lumira for a typical Information Consumer audience, but it is possible that information consumers will not be 100-percent satisfied with the tool and they may prefer a tool like SAP BusinessObjects Web Intelligence or a dashboard created with SAP BusinessObjects Design Studio for their work.

	Web Intelligence	Analysis Office	SAP Lumira	Design Studio & Design Studio Apps
Information Consumer	◕	◕	◑	◕
Business Analyst	◑	●	●	◑
Middle Management	◕	◕	◕	●
Decision Maker	◖	◔	◑	●

Figure 1.11 SAP BusinessObjects BI tools and personas

Figure 1.11 should not be seen as an exclusive statement, meaning that the user types can only use the tools shown. It is a guidance for tool selection. As you become more familiar with the tools in the SAP BusinessObjects BI portfolio, you will be able to use your own judgment and add your own criteria to the decision-making process. If a tool only covers part of a user type (for example, Web Intelligence and Business Analyst), you can assume that you will be able to address some of the requirements and needs of that particular user type with the tool, but that there still may be some areas that might be better addressed by a different tool. In addition, keep in mind when selecting the tool that there is no single tool that provides all the functionality that you might need. However, each tool does have a main purpose.

Going to the next step, we can also make a distinction between the pure consumption of content using the BI tools and the step of creating new content or editing existing content.

Figure 1.12 shows our user types and the mapping of the different BI clients to those user types for the consumption of the BI content. The BI tools are assigned in a priority order, so when it comes to your Decision Maker persona, your first choice for the consumption of BI content should be SAP BusinessObjects Design Studio.

CONSUME

Decision Maker	Middle Management	Analyst	Consumer
Design Studio	Design Studio	Analysis Office	Web Intelligence
Lumira	Analysis Office	Lumira	Analysis Office
Web Intelligence	Lumira	Design Studio	Lumira
	Web Intelligence	Web Intelligence	Design Studio

Figure 1.12 Consuming BI content

Figure 1.13 shows our user types with the assigned BI client tools according to the creation of new BI content or the workflow of editing existing content. As you will notice, this list of products assigned to each of the user types is smaller and more focused.

CREATE

Decision Maker	Middle Management	Analyst	Consumer
Design Studio Application	Analysis Office	Analysis Office	Web Intelligence
	Lumira	Lumira	Analysis Office
	Design Studio Application	Design Studio Application	
		Design Studio	

Figure 1.13 Creating BI content

In addition, you will notice that it does not show Design Studio but instead the Design Studio Application, meaning that this is an application that has been created using Design Studio for this user audience. It does not mean that we would expect, for example, our Decision Maker to go into Design Studio and create such an application or dashboard. In regards to Design Studio, the one user type of our four user types that we see creating BI content in Design Studio would be our Analyst/Power User user type.

In addition to the capabilities provided by the different BI client tools, you also need to keep in mind the skills required for your consumers and those users creating the content (creators). In Figure 1.14 you can see the BI products along with the level of skills required for the consumer.

Required Skills for the Consumer

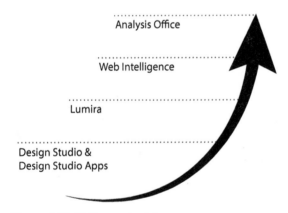

Analysis Office

Web Intelligence

Lumira

Design Studio &
Design Studio Apps

Figure 1.14 Skills required for consumers

As you can see here, the required skills are growing as the tool offers more self-service awareness and as analytical capabilities become a greater part of the self-service offering. The least amount of skills is required for the consumption of your Design Studio application as you can create those exactly tailored for your audience.

Figure 1.15 shows the required level of skills for those users who will create the BI content and, as you can see, the image shows a different level of skills compared to the pure consumption of the content. Here the products that offer a self-service focus require fewer skills as compared to those products that offer strong analytical and strong dashboarding types of capabilities.

Required Skills for the Creator

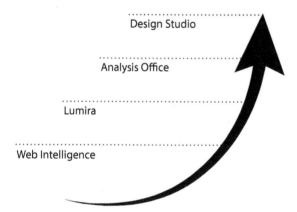

Design Studio

Analysis Office

Lumira

Web Intelligence

Figure 1.15 Skills required for creators

1.6 Summary

In this chapter, you learned about the criteria that are important to consider when selecting a tool from the SAP BusinessObjects BI portfolio. In addition, you learned that not only are the functional criteria important, but the user type and skillset are as well. In the next chapter, we will learn about the installation and configuration of the SAP BusinessObjects BI platform, and the important steps that need to be taken to integrate it with our SAP landscape.

CHAPTER 2

Installation and Configuration of SAP BusinessObjects BI 4.2

In this chapter we will focus on the installation and configuration of the SAP BusinessObjects BI platform as well as the SAP BusinessObjects BI client products, such as SAP BusinessObjects Lumira, SAP BusinessObjects Design Studio, SAP BusinessObjects Analysis, edition for Microsoft Office (Analysis Office), and Web Intelligence.

The installation and configuration of the SAP BusinessObjects BI platform server and client components consist of several main areas:

- Verifying the minimum versions of your SAP landscape.
- Installation of the server-side components of the SAP Business-Objects BI platform.
- Installation of the client-side components for the BI client tools.
- Post-installation steps for the SAP BusinessObjects BI platform.
- Configuring the SAP Authentication for the SAP BusinessObjects BI platform.
- Configuring the SAP HANA Authentication for the SAP Business-Objects BI platform.

In the following sections the installation and configuration are broken down into the installation of the server-side software and the client-side software. If you are following all these steps with a single hardware system available to you, you can install the server and client components from

SAP BusinessObjects on a single system. The recommended approach for this situation is to install the software in the following order:

1. SAP BusinessObjects BI platform

2. SAP BusinessObjects client tools

2.1 Verifying Your SAP Landscape

Before you start the installation and configuration of your SAP BusinessObjects system in combination with your SAP landscape, you should ensure that your SAP landscape is a so-called "supported platform."

You can review most recent supported platforms by going to the Platform Availability Matrix (PAM) at http://service.sap.com/pam. (Accessing the Service Marketplace requires a logon account that can be requested on the main page at http://service.sap.com.)

2.2 SAP BusinessObjects BI Platform— Server-Side Installation

In this section you'll install and configure the SAP BusinessObjects BI platform software. The focus here is to enable you to deploy a simple SAP BusinessObjects platform scenario. Further details on complex deployment scenarios and detailed installation material can be downloaded from http://service.sap.com/bosap-instguides.

Technical Prerequisites

Before you start the actual installation of the SAP BusinessObjects BI platform, ensure that the following requirements are met:

- Validate the exact details of the supported platforms and ensure that they match your environment. You can review the list of supported platforms at http://service.sap.com/pam.

- Check that your account for the operating system has administrative privileges.

- If you are planning to deploy on a distributed system, you need to have access to all machines via TCP/IP.

- You must have administrative access to the application server.

- You require access to a database system to install the system database for the SAP BusinessObjects BI platform or you can decide to use the default database system that is installed as part of the installation routine.

- SAP BusinessObjects Enterprise is 64-bit software, so ensure your server operating system is a 64-bit system.

- Windows .NET Framework version 3.5 Service Pack 01 is a prerequisite for installation on Windows operating systems.

- Windows Installation Program version 4.5 is a prerequisite for installation on Windows operating systems.

SAP BusinessObjects BI Platform—Default Installation

The default installation of SAP BusinessObjects Enterprise includes the Java application server Tomcat and Sybase SQL Anywhere for your system database.

For details on how to deploy the software using a different database system or a different supported application server including SAP J2EE, you can download the product documentation from http://help.sap.com.

SAP BusinessObjects Software Download

All available software from SAP BusinessObjects can be downloaded from the Service Marketplace by following the URL http://service.sap.com/swdc. In the category Installations and Upgrades you can then find the software listed alphabetically or you can use the option Software Downloads and navigate to the item By Category and select Analytics Solutions.

Installation Routine

For the installation routine we'll assume a single-server deployment scenario and we'll use Tomcat as the Java application server and Sybase SQL Anywhere for the system database. Based on the situation of performing a single-server deployment, the server name (BI4SERVER in our example) will also become the name for our SAP BusinessObjects BI platform system.

1. After you download the software you can start the installation routine by starting Setup.exe.

2. After you start the installation routine you are asked to select a language for the installation routine itself. This does not influence the language for the actual deployment of the software. In our example we'll select English as the setup language.

3. After accepting the setup language, you are presented with the Prerequisites check (see Figure 2.1) and you will not be able to install the SAP BusinessObjects BI platform unless all those listed prerequisites are met.

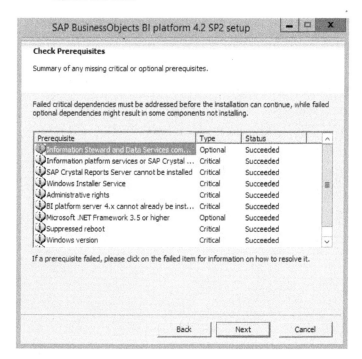

Figure 2.1 Check the prerequisites

4. After the prerequisites have been verified you are shown the start of the Installation Wizard (see Figure 2.2). Click Next to start the actual installation process.

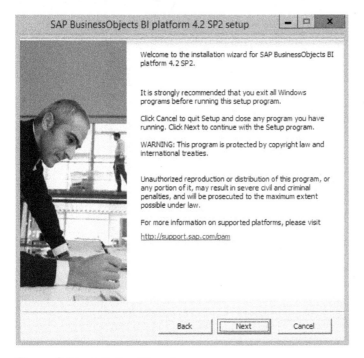

Figure 2.2 Installation Wizard

5. In the next two screens you are asked to accept the license agreement and enter the license keycode that you obtained. After this step, you have the choice of Language Packages (see Figure 2.3). This time the selection influences the availability of the software in different languages. For our installation we will select English.

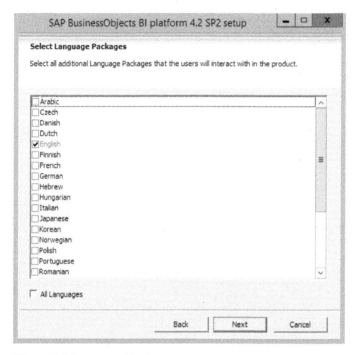

Figure 2.3 Language Packages

6. The next screen allows you to select the installation type. You can select between a Full installation, a Custom/Expand installation, or an installation option for the Web Tier (see Figure 2.4).

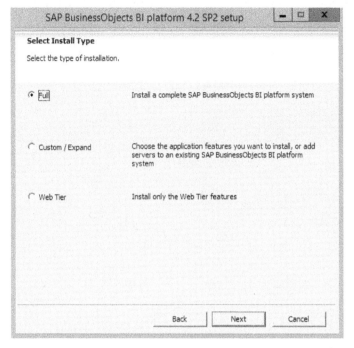

Figure 2.4 Installation type

In our example we are using the option Full so that we can install a complete SAP BusinessObjects BI platform system.

7. Next you can select the base folder for your SAP BusinessObjects
    Enterprise system. After configuring the folder location, you can
    decide which components you would like to install as part of your
    overall SAP BusinessObjects Enterprise system (see Figure 2.5).

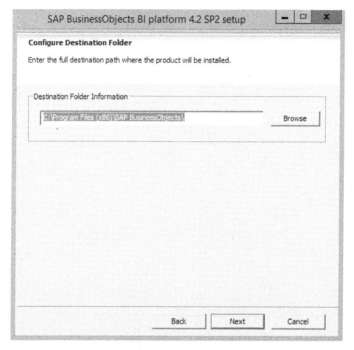

Figure 2.5 Configure the Destination Folder

8. After selecting the components, we are asked if we would like to install Sybase SQL Anywhere as our default system database (see Figure 2.6). For our example, select to install Sybase SQL Anywhere.

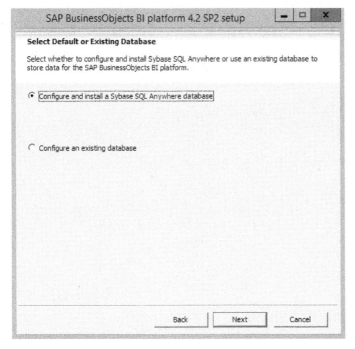

Figure 2.6 Select a default or existing database

CHAPTER 2

9. In the next step we are asked to either install the default Tomcat Java Web Application Server or to deploy the web applications manually later on (see Figure 2.7). In our example we will use the default Tomcat server.

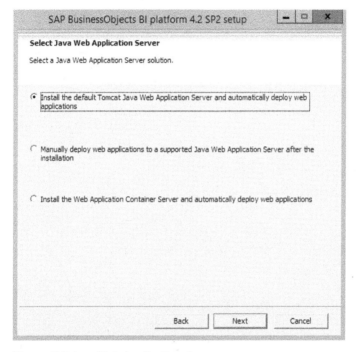

Figure 2.7 Java Web Application Server

CHAPTER 2

10. After configuring the installation of the Tomcat server, we are now asked if we would like to use Subversion for Version Management. In our example we will install Subversion (see Figure 2.8).

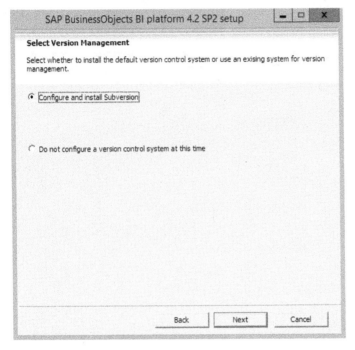

Figure 2.8 Version management

CHAPTER 2

11. Enter the name for the Server Intelligence Agent (SIA). In our example we will use the value BI4SERVER (see Figure 2.9).

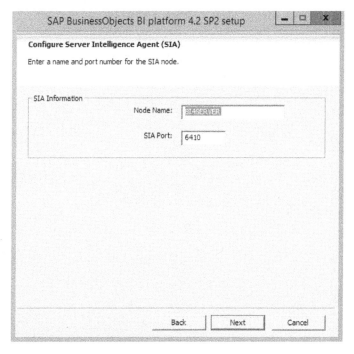

Figure 2.9 Configure the SIA

12. You are asked to configure the port for the Central Management Server (CMS). The default value is 6400 (see Figure 2.10).

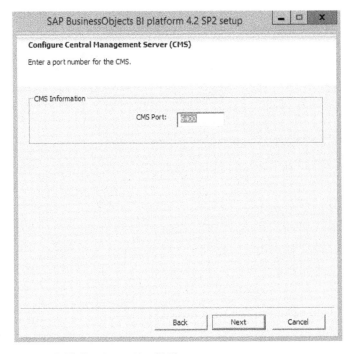

Figure 2.10 Configure the CMS

13. As part of the installation of our SAP BusinessObjects BI platform system, we also need to set up default passwords (see Figure 2.11). The installation routine creates an administrator account and you need to set up the default password.

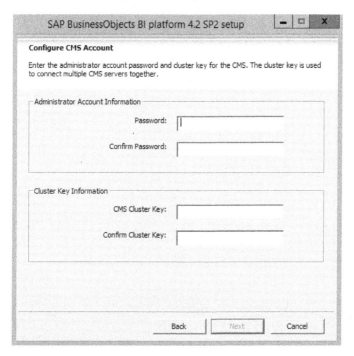

Figure 2.11 Default passwords

After configuring the default passwords for our administrator account we also need to set up the passwords for the Sybase SQL Anywhere installation (see Figure 2.12). The installation routine sets up an administrative account and a specific account that is being used by the SAP BusinessObjects BI platform system.

Password Requirements

As part of a new installation you are asked to enter the password for the administrator account as well as your underlying system database. The default configuration for passwords enforces a mixed-case password and requires you to use at least six characters. An example for a valid password would be Password1.

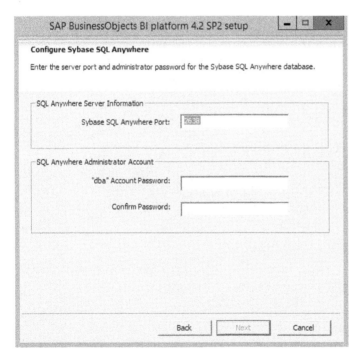

Figure 2.12 Sybase SQL Anywhere configuration

14. In the next step (see Figure 2.13) you need to configure the ports of the Java Application Server. In our example we selected to install Tomcat as part of the default installation. If you prefer to install with another Java Application Server, you have to manually deploy the applications.

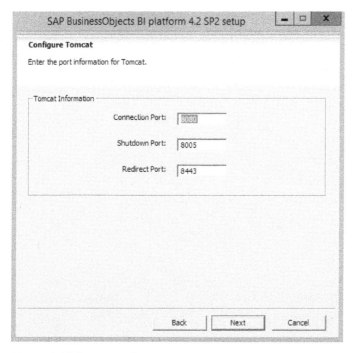

Figure 2.13 Configure Tomcat

In our example we accept the default ports for the Tomcat application server that the installation process recommends and continue to the next screen.

15. In the next step you need to configure the password for the Subversion repository as part of the Lifecycle management configuration.

16. In the next step you need to configure the port for the RESTful API for the SAP BusinessObjects BI platform (see Figure 2.14). In our example we will continue with the default port configuration.

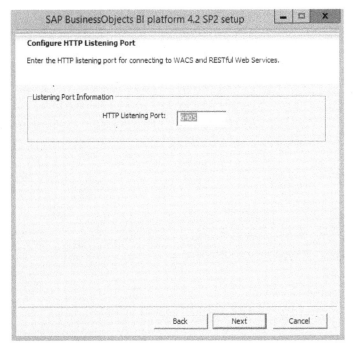

Figure 2.14 HTTP Listening Port

17. You are asked if you would like to configure the integration with Solution Manager (see Figure 2.15). We will use the option to not configure this at this point in time.

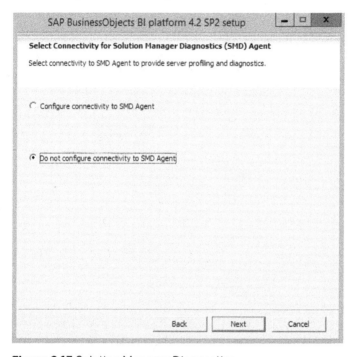

Figure 2.15 Solution Manager Diagnostics

18. As a next option you can configure integration into Introscope Enterprise Manager (see Figure 2.16) if you are not using Solution Manager for the tracing and monitoring of your SAP BusinessObjects BI platform system.

Figure 2.16 Introscope Enterprise Manager

CHAPTER 2

19. Finally, we can start the actual installation of our new SAP BusinessObjects BI platform system (see Figure 2.17). To do so, click Next.

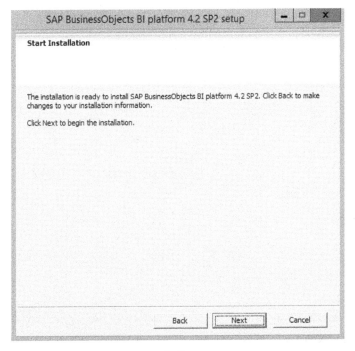

Figure 2.17 Start of the installation

20. After the installation routine has finished, you receive a note with some post-installation steps (see Figure 2.18).

Figure 2.18 Post-installation steps

You have now installed your first SAP BusinessObjects BI platform 4.2 environment and we can continue with the client part of the installation and configuration.

CHAPTER 2

2.3 SAP BusinessObjects BI Platform—Client Tools

In this section we'll cover the installation of the SAP BusinessObjects BI client tools. A more detailed technical overview of how those client tools can be used in combination with your SAP system will be presented in the following chapters.

Requirement for SAP GUI—Or Not?

SAP BusinessObjects Design Studio, Analysis Office, and SAP Lumira do not require a complete SAP GUI to be installed. Instead the file saplogon.ini with the system entries is already sufficient. Or you can use the alternative and configure the connection as part of the SAP BusinessObjects BI platform and always leverage the connections from your BI platform.

The following is a list of SAP BusinessObjects client tools that are delivered to you as part of the SAP BusinessObjects Client Tools installation:

- Web Intelligence Rich Client
- Report Conversion Tool
- Information Design Tool
- Translation Management Tool
- Query as a Web Service
- Universe Designer

Technical Prerequisites

- The account being used for the installation should have administrative rights on the system.
- Microsoft .NET Framework version 3.5 or higher needs to be installed on the client system.

Installation Routine

1. After you download the software you can start the installation routine by starting Setup.exe.

2. As the first part of the installation process for the SAP BusinessObjects client tools you select the language for the installation process.

3. After you select the language—in our example, English—you are presented with the check of the technical prerequisites (see Figure 2.19).

Figure 2.19 Prerequisites check

4. In the next steps you are shown the installation Welcome screen and you can accept the license agreement.

5. After accepting the license agreement, you can select the language packages for the SAP BusinessObjects client tools installation and you are asked to configure the destination folder.

6. As part of the next step you can select the components that should be installed (see Figure 2.20).

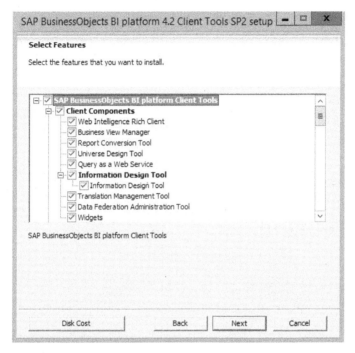

Figure 2.20 SAP BusinessObjects Client Tools

After you select the to-be installed components, you can start the actual installation process in the next step, and after a short while the SAP BusinessObjects Client Tools will be installed.

In this section we installed the SAP BusinessObjects BI client tools and in the next section we will install the client and server options from Analysis Office.

2.4 Analysis Office

As the next step of the client-side installation process, we will now install Analysis Office. It requires two installations: the actual client plug-in for Microsoft Office and the add-on for the SAP BusinessObjects BI platform.

Technical Prerequisites

- Microsoft Office 2010, 2013, or 2016 needs to be installed on the client system.

- Microsoft .NET Framework 4.5 Redistributable Package needs to be installed on the client system.

- Primary Interop Assemblies for Microsoft Office are required.

- The account being used for the installation should have administrative rights on the system.

Analysis Office Business Intelligence Platform Add-On Download

The software for the SAP BusinessObjects BI platform add-on for Analysis Office is not listed as part of the typical installations listing of software, but instead is only listed as part of the Support Packages and Patches area on the Service Marketplace.

Installation Routine

1. After you download the software you can start the installation routine.

2. In the first step you are presented with the Welcome screen for the installation of Analysis Office.

3. In the second step you can select the components for the installation (see Figure 2.21).

CHAPTER 2

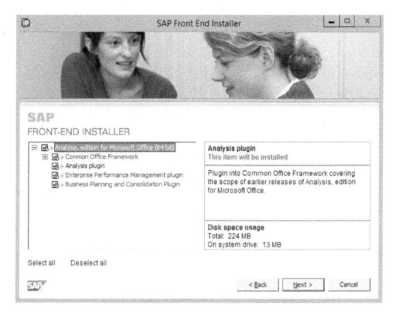

Figure 2.21 Component selection

4. Next you configure the destination folder for the installation (see Figure 2.22).

Figure 2.22 Destination folder

5. After these steps the actual installation process is started and you should be able to use Analysis Office shortly after this.

We installed the client part for Analysis Office, and now we continue with the installation of the add-on for the SAP BusinessObjects BI platform.

1. After you download the software you can start the installation routine.

2. You are asked to set a language for the installation routine. This language selection does not impact the actual installed software, but instead only changes the language for the installation routine itself.

3. In the next step you are presented with the prerequisite check (see Figure 2.23).

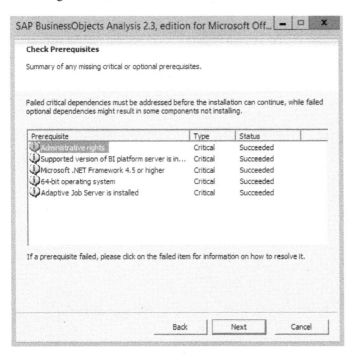

Figure 2.23 Prerequisites

4. You are presented with the Welcome screen and you have to accept the license agreement after that, so that the installation routine continues.

5. You are shown the folder location for the installation. Because this is an add-on installation, the folder location is already set based on the deployment and configuration of your SAP BusinessObjects BI platform system (see Figure 2.24).

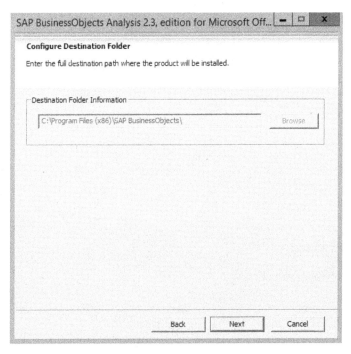

Figure 2.24 Folder location

6. You can select the components that will be installed (see Figure 2.25). In this example for the Analysis Office add-on there is only one component.

Figure 2.25 Component selection

7. After selecting the components, you are asked to provide the details to log on to your SAP BusinessObjects BI platform, so that the component can be deployed (see Figure 2.26).

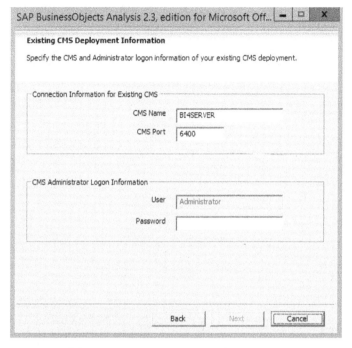

Figure 2.26 Deployment information

Next you can start the actual installation process, and after a short timeframe you should be able to leverage the add-on for Analysis Office.

In the next section we will install SAP BusinessObjects Design Studio as the client tool and the add-on components for the BI platform.

2.5 SAP BusinessObjects Design Studio

As the next step of the client-side installation process we will install SAP BusinessObjects Design Studio. The installation of SAP BusinessObjects Design Studio consists of two parts: the installation of the BI platform add-on and the installation of the SAP BusinessObjects Design Studio designer.

Technical Prerequisites

- The system should have a 64-bit operating system.
- The credentials used for the installation should have administrative rights on the system.
- Internet Explorer 9, 10, or 11 is required on the client system.
- The account being used for the installation should have administrative rights on the system.

Installation Routine

We will start the installation with the SAP BusinessObjects BI platform add-on:

1. After you download and extract the software fitting your environment, you can start the installation routine.

2. As the first part of the installation process you select the language for the installation process.

3. After you select the language—in our example, English—you are presented with the check of the technical prerequisites (see Figure 2.27).

Figure 2.27 Prerequisites check

4. You are shown the installation Welcome screen and you can accept the license agreement.

5. After accepting the license agreement, you can select the destination folder for the installation.

6. Next you can select the components that should be installed (see Figure 2.28). Table 2.1 shows the details for each of the server components.

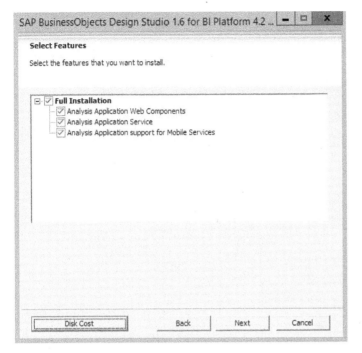

Figure 2.28 Component selection

Component	Description
Analysis Application Web Components	This feature makes it possible to display analysis applications in the BI launch pad, and use OpenDocument links to send direct links for analysis applications to other recipients. It also allows communication between the design tool and the BI platform, in order to save and launch analysis applications.
Analysis Application Service	This feature includes the Analysis Application Service running in the Adaptive Processing Server (APS). The Analysis Application Service makes it possible to execute analysis applications.
Analysis Application Support for Mobile Services	This feature adds mobile support to analysis applications, integrating analysis applications into SAP Mobile BI. It allows application users to access analysis applications on mobile devices.

Table 2.1 Server components

7. As we are deploying the BI platform add-on onto the existing SAP BusinessObjects BI platform, we need to provide the details for the CMS to log on to (see Figure 2.29).

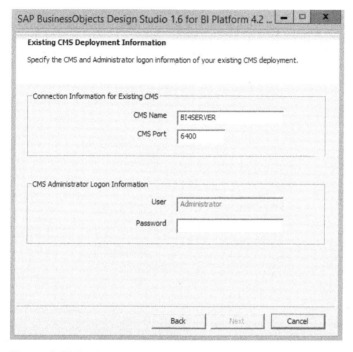

Figure 2.29 Deployment information

8. After step 7 the installation process is ready to start and you should have the SAP BusinessObjects Design Studio BI platform add-on installed in a few minutes.

We are now installing the SAP BusinessObjects Design Studio designer environment.

1. After you download and extract the software you can start the installation.

2. In the next step we select the component for SAP BusinessObjects Design Studio (see Figure 2.30).

Figure 2.30 Component selection

CHAPTER 2

3. After selecting the installation component, we can select the des-tination folder for SAP BusinessObjects Design Studio (see Figure 2.31).

Figure 2.31 Destination folder

4. After this step the installation process is ready to start and you can use SAP BusinessObjects Design Studio after a few minutes.

In this section we installed SAP BusinessObjects Design Studio on your client system as well as part of your SAP BusinessObjects BI plat-form. In the next section we will install the SAP Lumira Desktop and the BI platform add-on.

2.6 SAP Lumira

As the last step in our client-side installation process, we are going to install SAP Lumira Desktop and the SAP BusinessObjects BI platform add-on for SAP Lumira.

Technical Prerequisites

- The account being used for the installation should have administrative rights on the system.

- You will require at least SAP BusinessObjects BI Platform 4.1 Support Package 05 or higher to leverage the SAP Lumira server for the SAP BusinessObjects BI platform.

Installation Routine

We will start the installation with the SAP BusinessObjects BI platform add-on for SAP Lumira:

1. After you download and extract the software fitting your environment, you can start the installation routine.

2. Select the language for the installation process.

3. Accept the license agreement (see Figure 2.32).

Figure 2.32 License Agreement

4. After accepting the license agreement, you can choose the to-be installed components (see Figure 2.33).

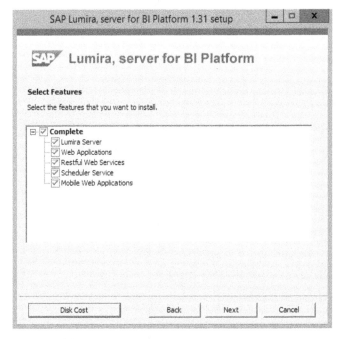

Figure 2.33 Selected components

5. You are asked to provide the details of your CMS (BI4SERVER for our example) and you have to provide the details for the administrator credentials to log on to your SAP BusinessObjects BI platform system (see Figure 2.34).

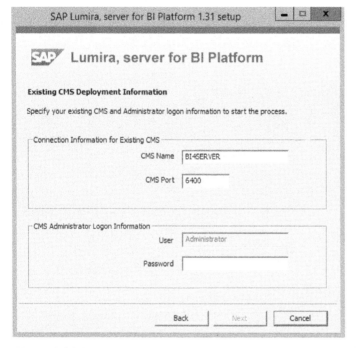

Figure 2.34 Deployment information

6. After this step, the installation routine starts and you should be able to view your SAP Lumira documents via your SAP BusinessObjects BI platform shortly after this.

After installing the server components for SAP Lumira, we will now continue with the installation of SAP Lumira Desktop.

1. After you download and extract the software fitting your environment, you can start the installation routine.

2. As the first part of the installation process you need to configure the destination folder for the installation (see Figure 2.35).

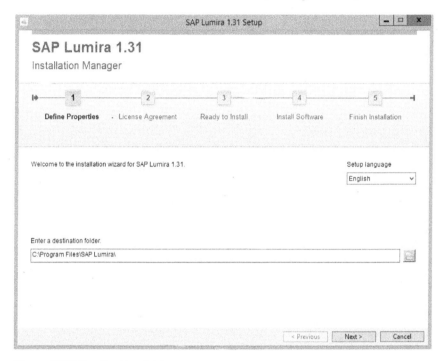

Figure 2.35 Destination folder

3. After you configure the folder location, you are asked to accept the license agreement in the next screen (see Figure 2.36).

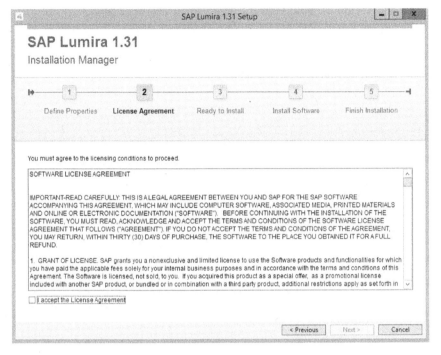

Figure 2.36 License Agreement

CHAPTER 2

4. After accepting the license agreement, the installation routine is ready to get started (see Figure 2.37). Click Next.

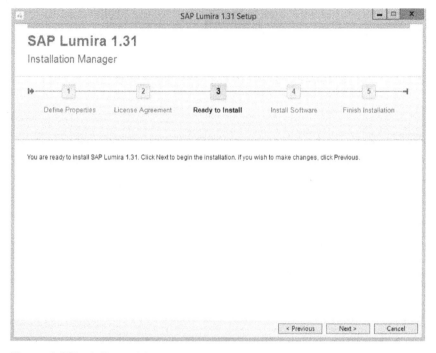

Figure 2.37 Installation Manager

After a short period of time, you are asked if you would like to start SAP Lumira Desktop and the product is ready to be used.

2.7 Post-Installation Steps

After completing the installation steps, we have to ensure that several post-installation steps are performed for each of the BI client products. In this section we will outline each of those steps product by product.

2.7.1 SAP BusinessObjects Design Studio— Post-Installation Steps

After the installation and configuration of SAP BusinessObjects Design Studio server components and SAP BusinessObjects Design Studio designer, the following are some tasks you want to ensure are completed next:

- Check the status of the Analysis Application Service.
- Configure the maximum number of client sessions.
- Create a user group for working with SAP BusinessObjects Design Studio dashboards.
- Define your Design Studio administrators.
- Configure parameters for your Analysis Application Service.
- Configure the mobile usage for your dashboards.
- Configure the number of sessions for parallel query execution.

Check the Status of the Analysis Application Service

After you install the new component as part of your SAP BusinessObjects BI platform, you should check if the Analysis Application Service has been added to your landscape and is up and running.

1. Log on to the Central Management Console (CMC) of your SAP BusinessObjects BI platform with an administrative account.

2. On the Home page navigate to the area Servers (see Figure 2.38).

Figure 2.38 Area Servers

3. Select the category Analysis Services in the list of Service Categories (see Figure 2.39).

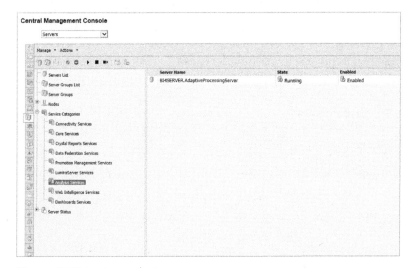

Figure 2.39 Analysis services

4. Double-click the entry for the Adaptive Processing Server to open the Properties (see Figure 2.40).

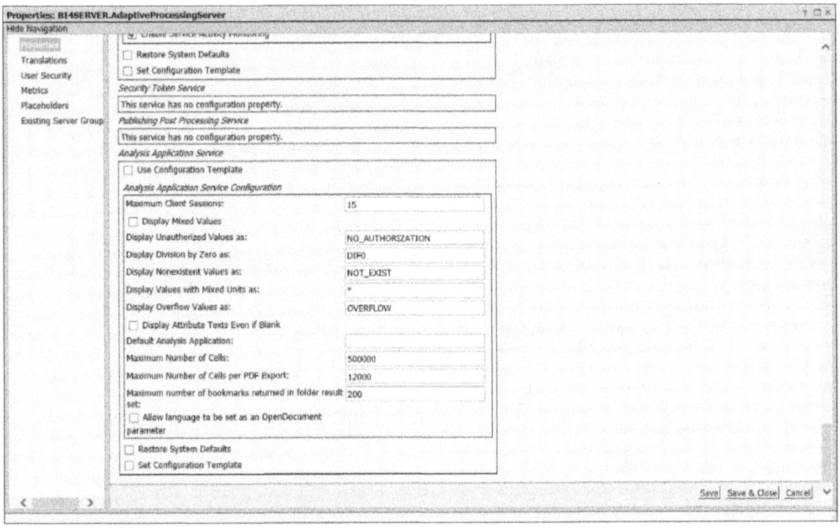

Figure 2.40 Server properties

5. Scroll down to the Analysis Application Service area of the properties.

In case the initial start of the service failed, you receive a message here with a log ID. The actual logfile can be found in the logging directory, on a Windows platform by default: \<Installation Directory>\SAP BusinessObjects Enterprise XI 4.0\logging.

Configure the Maximum Number of Client Sessions
As a next step you want to ensure that you configure the maximum number of client sessions for your SAP BusinessObjects Design Studio service so that it fits your needs for the overall landscape.

1. Log on to the CMC of your SAP BusinessObjects BI platform with an administrative account.

2. On the Home page navigate to the area Servers (see Figure 2.38).

3. Select the category Analysis Services in the list of Service Categories (see Figure 2.39).

4. Double-click the entry for the Adaptive Processing Server to open the Properties (see Figure 2.40).

5. Scroll down to Analysis Application Service area of the properties.

6. You can now change the value for the property Maximum Client Sessions.

Create a User Group for Working with SAP BusinessObjects Design Studio Dashboards

With the SAP BusinessObjects BI platform you can also set up users and user groups. You can assign specific rights and permissions to define which kind of content each user is able to consume and which type of functionalities each user is able to leverage from the SAP BusinessObjects BI platform. For users to be able to consume content created with SAP BusinessObjects Design Studio, you first have to create a new user group and assign the necessary security level.

1. Log on to the CMC of your SAP BusinessObjects BI platform with an administrative account.

2. On the Home page navigate to the area Users and Groups (see Figure 2.41).

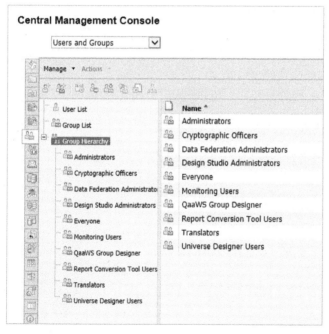

Figure 2.41 Users and Groups

3. Follow menu Manage > New > New Group.

4. Enter Design Studio Users as the Group Name.

5. Click OK.

6. In the list of areas for the CMC select the entry Applications (see Figure 2.42).

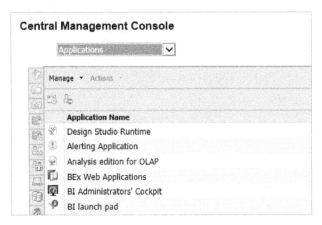

Figure 2.42 Select Applications

7. Select the entry Design Studio Runtime.

8. Use a right-click and select the entry User Security (see Figure 2.43).

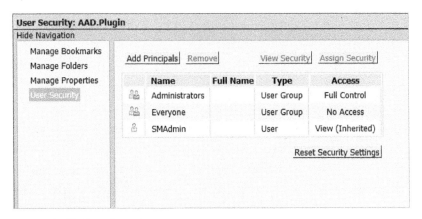

Figure 2.43 User Security

9. Click Add Principals.

10. Select the previously created User Group.

11. Add the User Group to the list of Selected Users or Groups.

12. Click Add and Assign Security.

13. Select the Access Level View and add the Access Level to the list of Assigned Access Levels (see Figure 2.44).

14. Click OK.

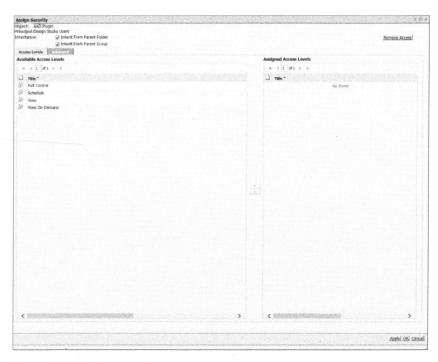

Figure 2.44 Assign security

You should now be able to add your users to the newly created user group. Users will inherit the assigned security. In addition to the rights to work with Design Studio, you also need to ensure that your users are given access to the required data connections.

Define Your Design Studio Administrators

As part of the installation of the SAP BusinessObjects Design Studio add-on to the SAP BusinessObjects BI platform, you also will receive a new user group—Design Studio Administrators. Only members of the Design Studio Administrator group or members of the BI Platform Administrators group are able to install extensions for SAP BusinessObjects Design Studio.

To add new members to the Design Studio Administrators, follow these steps:

1. Log on to the CMC of your SAP BusinessObjects BI platform with an administrative account.

2. On the Home page navigate to the area Users and Groups.

3. Select the entry Group List.

4. Select the entry Design Studio Administrators.

5. Use a right-click and select the menu Add Members to Group (see Figure 2.45).

CHAPTER 2

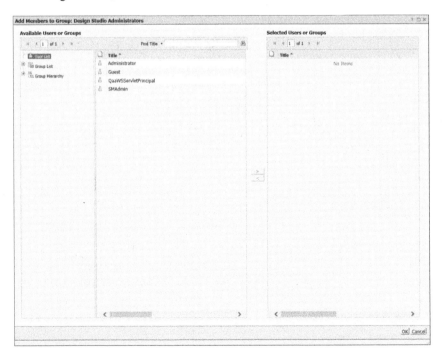

Figure 2.45 Add members

6. Add the members required to the list of Selected Users or Groups.

7. Click OK.

You can now add your users to the Design Studio Administrators group and those members will be able to install extensions.

Configure Parameters for Your Analysis Application Service

In addition to the previously configured settings, you can configure several options that have an impact on how specific content is displayed in all your dashboards using SAP BusinessObjects Design Studio.

To configure these parameters, follow these steps:

1. Log on to the CMC of your SAP BusinessObjects BI platform with an administrative account.

2. On the Home page navigate to the area Servers.

3. Select the category Analysis Services in the list of Service Categories.

4. Double-click the entry for the Adaptive Processing Server to open the Properties.

5. Scroll down to the Analysis Application Service area of the properties (see Figure 2.46).

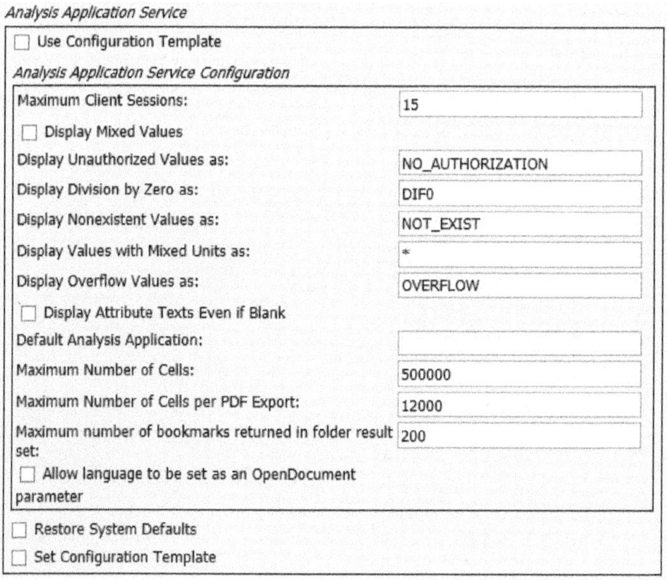

Figure 2.46 Analysis Application Service configuration

6. You can follow the details in Table 2.2 to configure these parameters.

Parameter	Description
Display Mixed Values	This option allows you to enable or disable the display of mixed unit/currency values.
Display Unauthorized Values as	Here you can enter a string value, which will be displayed for the values, for which the user is not authorized.
Display Division by Zero as	Here you can enter a string value, which will be displayed for division by zero values.
Display Nonexistent Values as	Here you can enter a string value, which will be displayed for values, which cannot be determined.
Display Values with Mixed Units as	Here you can enter a string value, which will be displayed for values, which consist of mixed currencies and units.
Display Overflow Values as	Here you can enter a string value, which will be displayed for values with a numerical overflow.
Display Attribute Texts Even if Blank	Display attribute text when the check box is selected, even if attribute text is blank
Maximum Number of Cells	Here you can specify the maximum number of cells per data source. If the result set for a data source exceeds the limit set, then no data will be returned, and a warning is displayed in the crosstab.
Maximum Number of Cells per PDF Export	Here you can specify the maximum number of cells that can be exported to PDF from within an analysis application.

Table 2.2 Analysis Application Service configuration

7. After making the necessary changes, click Save & Close.

We have now configured the detailed parameters for our Analysis Application Service and in the next step we will configure mobile usage in combination with the SAP BusinessObjects BI platform.

Configuring Mobile Usage

In case you would like to share your dashboards created with SAP BusinessObjects Design Studio via mobile devices, you have to configure the mobile usage as part of your SAP BusinessObjects BI platform.

Follow these steps to configure the mobile usage:

1. Log on to the CMC of your SAP BusinessObjects BI platform with an administrative account.

2. On the Home page navigate to the area Categories.

3. Follow menu Manage > New > New Category.

4. Enter Mobile as new category name (see Figure 2.47).

Figure 2.47 Categories

5. Now navigate to the area Applications in the CMC.

6. Select the application SAP BusinessObjects Mobile.

7. Use a right-click and select the menu Properties (see Figure 2.48).

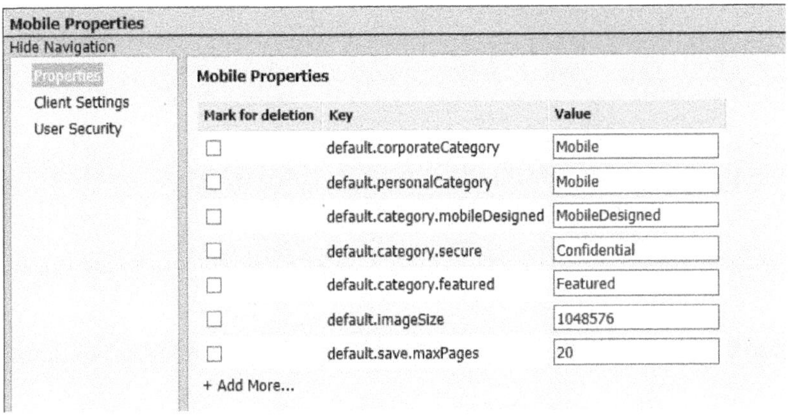

Mobile Properties

Hide Navigation

Properties
Client Settings
User Security

Mobile Properties

Mark for deletion	Key	Value
☐	default.corporateCategory	Mobile
☐	default.personalCategory	Mobile
☐	default.category.mobileDesigned	MobileDesigned
☐	default.category.secure	Confidential
☐	default.category.featured	Featured
☐	default.imageSize	1048576
☐	default.save.maxPages	20

+ Add More...

Figure 2.48 Application Properties

8. You can now enter your newly created category as the default category for your mobile usage.

9. After making the necessary changes, click Save & Close.

CHAPTER 2

For your dashboard to leverage the category for mobile usage, you have to manually assign the category to your dashboard.

1. Log on to the BI launch pad of your SAP BusinessObjects BI platform.

2. In the BI launch pad navigate to the folder where you stored your SAP BusinessObjects Design Studio dashboard.

3. Select the dashboard.

4. Use a right-click and select the menu Categories.

5. Now assign the category for mobile usage to your dashboard.

6. Click OK.

After configuring the mobile category and assigning the category to your dashboard, you can now view the dashboard using the SAP Mobile BI application on your mobile device.

Configuring Parallel Query Execution

Since release 1.5 of SAP BusinessObjects Design Studio, it is possible to run queries for the used data sources in your dashboard in parallel instead of in sequence. As parallel query execution does consume additional CPU resources and also generates an additional session in the back end, you should carefully limit the maximum number of sessions allowed for the parallel query execution.

To configure a limit, follow these steps:

1. Log on to the CMC of your SAP BusinessObjects BI platform with an administrative account.

2. On the Home page navigate to the area Servers.

3. Navigate to the entry Service Categories.

4. Select the entry Analysis Services.

5. Double-click the entry for the Adaptive Processing Server to open the Properties (see Figure 2.49).

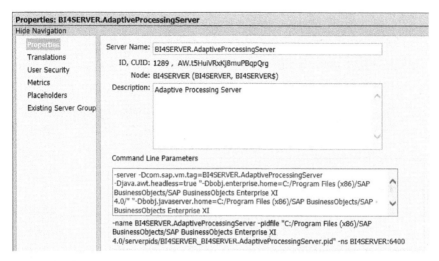

Figure 2.49 Properties

6. You can add the following command line parameter:

   -DAAD_GROUPS_THREADCOUNT=<value>

7. Replace the placeholder <value> with a numeric value. The default value is 500.

8. Add the command line parameter to the already existing Command Line.

9. Click Save & Close.

10. Select the Adaptive Processing Server.

11. Use a right-click and select the menu Restart Server.

In this section we reviewed all the post-installation steps for SAP BusinessObjects Design Studio. In the next section we will look at the post-installation steps for Analysis Office.

2.7.2 Analysis Office—Post-Installation Steps

After the installation and configuration of Analysis Office, there are two important tasks you should consider:

- Configuring the load behavior.
- Configuring the connection to your SAP BusinessObjects BI platform.

Configuring the Load Behavior

Analysis Office is a plug-in for Microsoft Excel and Microsoft Power-Point. Unless you prefer to always explicitly call the menu items for SAP BusinessObjects Analysis for Microsoft Excel and SAP BusinessObjects Analysis for Microsoft PowerPoint, you should consider configuring the load behavior of the necessary plug-ins so that the plug-ins are always loaded when Microsoft Excel or Microsoft PowerPoint are started.

To configure the load behavior, follow these steps:

1. Open the registry editor on your client system where you installed Analysis Office.

2. Navigate to the folder HKEY_CURRENT_USER\Software\Microsoft\Office\Excel\Addins\SapExcelAddIn.

3. Select the entry LoadBehavior.

4. Configure the parameter according to the values shown below:

   - Value 0: The Add-In is disabled. Users have the option to temporarily enable the Add-In.
   - Value 1: The Add-In is enabled, but on the next restart the Add-In will be disabled.
   - Value 2: The Add-In is disabled and the user can enable it. In case users enable the Add-In, the value for the Load Behavior parameter changes to 3.
   - Value 3: The Add-In is enabled.

5. Close the registry.

Analysis Office should now be loaded according to your configuration.

Configuring the Connection to Your SAP BusinessObjects BI System

Your users will be able to log on to the SAP BusinessObjects BI platform system and in that way leverage all the connections defined on the BI platform. In case your users only leverage a single SAP BusinessObjects BI platform system, they will be easily able to configure it as part of Analysis Office. However, if you prefer to provide your users with a predefined list of systems you can follow the steps outlined below.

The suggested settings outlined below can be done as an administrator or as a user, which has implications for the location of the settings.

- As an administrator, you maintain the settings in the file Ao_app. config. The file is located in the file system under C:\ProgramData\ SAP\Cof.

- As a user, you can change the settings in the file system under Users\<UserID>\AppData\Roaming\SAP\Cof. The file name for changing the settings is ao_user_roaming.config.

To configure a predefined SAP BusinessObjects BI system:

1. Navigate to the folder location C:\ProgramData\SAP\Cof on your system with Analysis Office installed.

2. Open the file Ao_app.config.

3. Navigate to the setting <BOESystems> in the file below the <configSections>.

4. Add the necessary details. See the sample XML file in Figure 2.50.

```
<BOESystems><![CDATA[
  <?xml version="1.0" encoding="utf-16"?>
    <ArrayOfCoBoeSystemInfo xmlns:xsi="http://www.w3.org/2001/
    XMLSchema-instance" xmlns:xsd="http://www.w3.org/2001/
    XMLSchema">
      <CoBoeSystemInfo>
        <SystemId>%SYSTEM ID%</SystemId>
        <SystemName>%SYSTEM NAME%</SystemName>
        <Hostname>%HOSTNAME%</Hostname>
        <Scheme>%PROTOCOL%</Scheme>
        <Port>%PORT%</Port>
        <SessionServiceUrl>/dswsbobje/services/Session
        </SessionServiceUrl>
        <Active>True</Active>
        <UseSso>True</UseSso>
        <LastUsedAuthentication>%AUTHENTICATION%
        </LastUsedAuthentication>
        <CMSNames><string><%CMS NAME %></string></CMSNames>
      </CoBoeSystemInfo>
    </ArrayOfCoBoeSystemInfo>
]]></BOESystems>
```

Figure 2.50 Sample XML

Table 2.3 explains the placeholders used in the XML.

Placeholder	Description
< SYSTEM ID >	Enter a name for your system.
< SYSTEM NAME >	Enter a name for your system.
< HOSTNAME >	Enter the hostname of the application server of your SAP BusinessObjects Enterprise system.
< PROTOCOL >	Enter the protocol you are going to use. HTTP or HTTS.
< PORT >	Enter the port of your application server.
< CMS NAME >	Enter the name of your CMS of your SAP BusinessObjects Enterprise system.
< AUTHENTICATION >	Specify the authentication type for the logon.

Table 2.3 Connection details

- By setting the value True for the option <Active>, the configured connection becomes the default connection for your user.

5. Save the changes to the file and ensure that you are starting Analysis Office, fresh.

Multiple SAP BusinessObjects BI Platform Systems

Both options—a manual entry or the XML file—provide you the option to add multiple SAP BusinessObjects BI platform system definitions to the list. In case you define multiple systems the user will have the option to select one system from the list of available entries.

In this section we reviewed the options to configure the connection to your SAP BusinessObjects BI platform system for Analysis Office, as well as the load behavior for the Microsoft Office Add-In. In the next section we will review the post-installation steps for SAP Lumira.

2.7.3 SAP Lumira—Post-Installation Steps

After installing the SAP Lumira Desktop and SAP Lumira, server for SAP BusinessObjects BI platform, you should look at the configuration of the maximum data volume for the Visualize room in SAP Lumira and adjust the limit according to your requirements.

Configure SAP Lumira Server Services Properties

You can adjust the maximum limit for the data volume that is supported by the SAP Lumira Viewer for the Visualize room.

To adjust the limit, follow these steps:

1. Log on to the CMC of your SAP BusinessObjects BI platform with an administrative account.

2. On the Home page navigate to the area Servers.

3. In the Service Categories select the entry Lumira Server Services.

4. Select the entry for the SAP Lumira server.

CHAPTER 2

5. Use a right-click and select the menu Properties (see Figure 2.51).

Figure 2.51 Server properties

6. Add the following command line to the already existing command line parameters:

   -Dhilo.maxvizdatasetsize=<value>

   Replace the placeholder <value> with the limit for the amount of data cells you would like to configure.

7. Click Save & Close.

8. Select the entry for the SAP Lumira server.

9. Use a right-click and select the menu Restart Server.

In this section we reviewed the post-installation steps for SAP Lumira. In the next section we will configure the SAP Authentication for our SAP BusinessObjects BI platform.

2.8 Configuring SAP Authentication

After installing the SAP BusinessObjects BI platform you need to configure the SAP Authentication so that you can import the SAP users and roles into your SAP BusinessObjects BI platform system and allow them to use their SAP credentials to get access to their BI assets. In addition to setting up the SAP Authentication, we will also configure the trust between the SAP system and your SAP BusinessObjects BI system, so that Single Sign-On (SSO) for BI clients such as Analysis Office and SAP BusinessObjects Design Studio will work.

Technical Prerequisites

To be able to use SSO between the SAP system and your SAP Business-Objects BI platform, you need to configure the SAP system to accept SSO logon tickets and to create them. This involves setting parameter values in the profile of your SAP system via transaction code RZ10; setting or changing those values requires a restart of the system (see Table 2.4).

Profile Parameter	Value	Comment
login/create_sso2_ticket	1 or 2	Use the value 1 if the server possesses a public key certificate signed by the SAP CA (SAP Certification Authority). Use the value 2 if the certificate is self-signed. If you are not sure, then use the value 2.
login/accept_sso2_ticket	1	Use the value 1 so that the system will also accept logon tickets.

Table 2.4 Single Sign-On Profile Parameters

Configuring SAP Authentication

To configure the SAP Authentication, follow these steps:

1. To start the configuration of the SAP Authentication you need to log on to the CMC with an administrative account.

2. After a successful authentication you'll be presented with the main screen of the CMC. To configure the SAP Authentication details, select the option Authentication on the main page, and the system then presents the list of available authentication providers (see Figure 2.52).

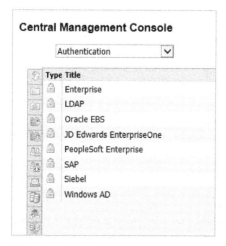

Figure 2.52 Authentication provider

3. You can now double-click the SAP authentication provider, and the system presents you with the screen shown in Figure 2.53 so that you can create a new SAP system entry.

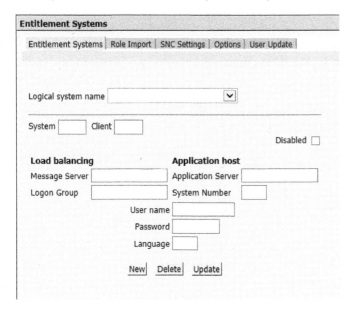

Figure 2.53 Entitlement Systems

4. In this first step of the configuration you can identify the SAP systems that you want to use in combination with the SAP BusinessObjects BI platform. Only systems that are configured during this step can use the full functionality of the SAP BusinessObjects landscape.

5. The configuration screen allows you to configure your system based on an Application Server with a System Number or a Message Server with a Logon Group. Start by entering the System ID of your SAP system and the Client Number; then enter either a combination of message server with a logon group or application server with a system number. When you're finished with the configuration, your system will appear in the Logical system name field so that you can select any of the configured systems later on.

6. The user account that is requested here is used only for administrative tasks, such as reading the users and roles from the SAP system or validating role membership of users authenticating against the system. The user account requires only a bare minimum of authorizations on the SAP side. Enter the user credentials from the SAP account that you created based on those authorizations and the language code, and click the Update button to save your entries.

Table 2.5 lists all necessary authorization objects and authorization values that need to be assigned to the user credentials that are being leveraged in the SAP Authentication configuration dialog.

Authorization Object	Authorization Field	Value
S_DATASET	ACTVT	33,34
	FILENAME	*
	PROGRAM	*
S_RFC	ACTVT	16
	RFC_NAME	BDCH, STPA, SUSO, SUUS, SU_USER, SYST, SUNI, PRGN_J2EE, /CRYSTAL/SECURITY
	RFC_TYPE	FUGR
S_USER_GROUP	ACTVT	3
	CLASS	* For security reasons you can also list user groups that have access to SAP BusinessObjects Enterprise.

Table 2.5 Authorizations for SAP Authentication configuration

7. Navigate to the Options tab of the SAP Authentication (see Figure
   2.54). This dialog allows you to configure the behavior of the SAP
   Authentication for your SAP BusinessObjects Enterprise server. All
   configurations you perform in this step will apply to all the SAP
   entitlement systems. A detailed explanation of all possible options
   is shown in Table 2.6.

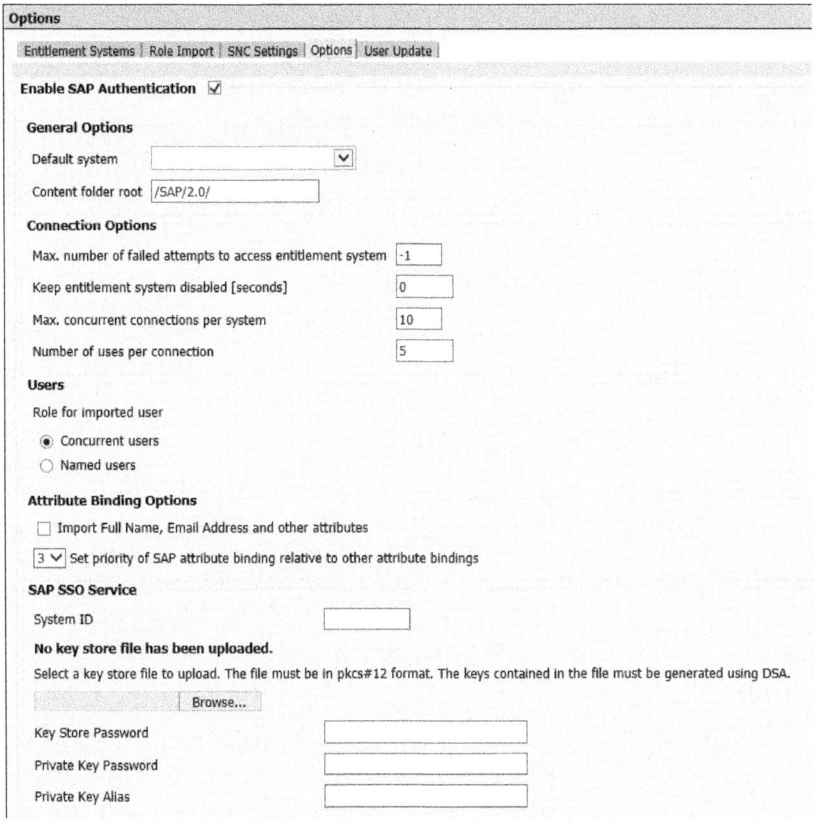

Figure 2.54 Authentication options

CHAPTER 2

Configuration Option	Description
Enable SAP Authentication	You can use this check box to disable SAP Authentication for the SAP BusinessObjects system. If you want to disable only a single entitlement SAP system, you can do that on the Entitlement Systems tab.
Default System	The default system is used when a user is trying to authenticate SAP credentials without specifying the SAP system. A common scenario is a user navigating from the SAP Enterprise Portal to the SAP BusinessObjects BI platform system without specifying for which SAP system the authentication should be performed. In such a scenario the default system will be used as a fallback, and the SAP BusinessObjects BI platform system will try to authenticate the user against the default system.
Max number of failed attempts to access entitlement system	You can use this setting to configure how many attempts the SAP BusinessObjects server should make to connect to an SAP system that is temporarily not available. The value -1 is used for an unlimited number, the value 0 is used for one attempt, and values larger than 0 represent the actual numeric value.
Keep entitlement system disabled (seconds)	This setting is used in combination with the option above to configure the time in seconds that the SAP BusinessObjects BI platform server will wait before trying to access an SAP system again that previously had reached the maximum number of failed attempts.
Max concurrent connections per system	Here you can configure the maximum number of concurrent connections that SAP BusinessObjects BI platform can keep open towards the SAP entitlement system.
Number of uses per connection	Here you can specify the number of operations that can be performed on a single connection. For example, setting the value to 5 will result in the connection being closed after 5 operations or logons on this connection.
Role for imported users	This option allows you to configure if the imported SAP users should be marked for a concurrent or named user license.
Import Full Name and Email Address	Select this option if you want to import the full names and descriptions used in the SAP accounts into the user objects in the SAP BusinessObjects BI platform.
Set priority of SAP attribute binding relative to other attributes binding.	Specifies a priority for binding SAP user attributes. The option can be set to a value between 1 and 3. Value 1 = SAP attributes take priority in scenarios where SAP and other plug-ins (Windows AD and LDAP) are enabled. Value 3 = Attributes from other enabled plug-ins will take priority

Table 2.6 Configuration options

We will discuss the SSO Service shown as part of the Options tab and the necessary configuration steps in a later section as part of the configuration of SSO Trust between our SAP system and our SAP BusinessObjects BI platform.

8. After configuring the options for the SAP Authentication, you can navigate to the Role Import tab (see Figure 2.55) to use the SAP roles and users for your SAP BusinessObjects BI platform. The screen allows you to select one system from the list of SAP entitlement systems you created and provides a list of available SAP roles. Each of these roles can be imported by adding it to the Imported Roles field. As soon as you click the Update button, each of the SAP roles will become a user group in your SAP BusinessObjects system and, if you configured it to automatically import the users, the assigned users will become users in your system.

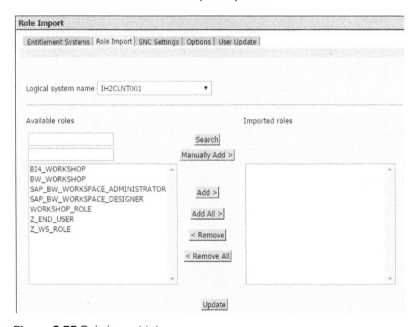

Figure 2.55 Role Import tab

For each of the imported SAP roles, the system generates an SAP BusinessObjects user group based on the following logic:

[SAP System ID]~[SAP client number]@[SAP role]

Example:

IH1~800@SAP_BUSINESSOBJECTS_CONTENT_ROLE

Each imported user follows the syntax:

[SAP System ID]~[SAP client number]/[SAP user]

Example:

IH1~800/DEMO_USER

Role Import and Authentication with SAP Credentials

The ROLE IMPORT option only offers those SAP roles that have users assigned to them. SAP roles with no user assignment will not be shown as available roles for the import. After you import the SAP roles, each of the generated user groups has no assigned rights in the SAP BusinessObjects system. Those resulting user groups are only assigned to the standard SAP BusinessObjects user group Everyone.

After you import the SAP roles and users from your entitlement system, you should be able to use your SAP credentials and authenticate against the SAP BusinessObjects BI platform and log on with those credentials to the BI launch pad or the CMC using SAP Authentication.

9. After importing the roles navigate to the User Update tab (see Figure 2.56).

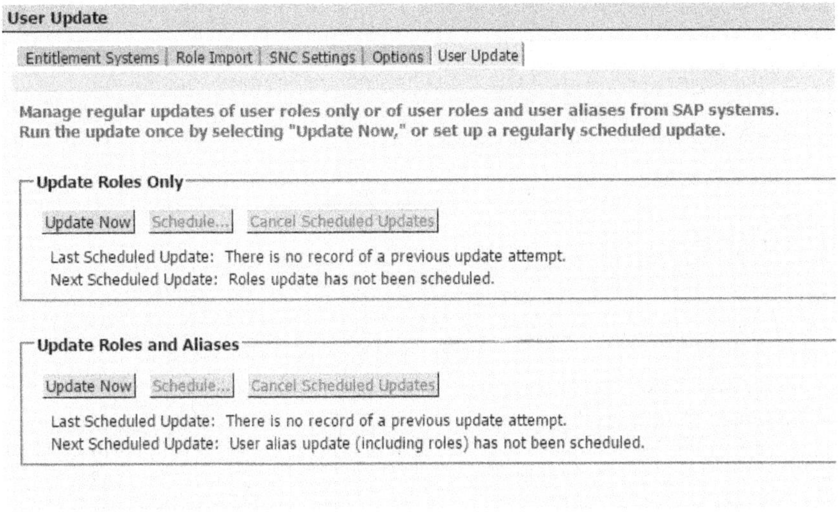

User Update

Entitlement Systems | Role Import | SNC Settings | Options | User Update

Manage regular updates of user roles only or of user roles and user aliases from SAP systems. Run the update once by selecting "Update Now," or set up a regularly scheduled update.

Update Roles Only

[Update Now] [Schedule...] [Cancel Scheduled Updates]

Last Scheduled Update: There is no record of a previous update attempt.
Next Scheduled Update: Roles update has not been scheduled.

Update Roles and Aliases

[Update Now] [Schedule...] [Cancel Scheduled Updates]

Last Scheduled Update: There is no record of a previous update attempt.
Next Scheduled Update: User alias update (including roles) has not been scheduled.

Figure 2.56 User Update

Here you can set up a regular update of the imported SAP roles and users by simply setting up a schedule.

Configuring the SSO Token Service

In addition to setting up the SAP Authentication, you also have to configure the trust between your SAP system and your SAP BusinessObjects BI system by leveraging the SSO Token Service. Starting with SAP BusinessObjects 4.0, the SAP BusinessObjects BI platform provides you with an SSO Token Service that is capable of generating Assertion Tickets. The generated Assertion Tickets can help you to achieve an SSO to the SAP system in several scenarios:

- You can leverage this new service to enable scheduling of BI content with SSO.

- You can leverage this new service to enable publications (report bursting) of BI content with SSO.

- You can leverage this new service to enable SSO to the SAP system for cases where multiple user aliases are involved.

- You can leverage this new service to enable SSO for scenarios in which SAP credentials are not involved as part of the initial user authentication; for example, using Windows Active Directory credentials to authenticate against the SAP BusinessObjects Enterprise system and still achieve SSO towards the SAP system.

- You can leverage this service to achieve SSO for thick-client products such as Analysis Office.

Here is an example in which you would be using the SSO Token Service:

You could have an SAP BusinessObjects BI platform user "User A" with a user alias configured. The alias is the SAP account for "User A"— let's assume "SAP User A" for now—and the user will then be able to log on with the Enterprise authentication and the credentials "User A" to the SAP BusinessObjects BI platform system, or to a thick client such as Analysis Office, and achieve SSO to the SAP system.

The configuration for the SSO Token Service includes several steps:

- Generating a key store file and certificate for the SAP BusinessObjects Enterprise system.

- Importing the certificate to the SAP system.

- Setting up SSO in the SAP BusinessObjects Enterprise CMC.

- Adding the SSO Token Service to your SAP BusinessObjects Enterprise system.

In the next steps we will configure the SSO Token Service as part of the SAP BusinessObjects Enterprise system.

Generating a Key Store File and Certificate

As the first step for the overall configuration you need to generate a key store file and a certificate for your SAP BusinessObjects BI platform system. To generate the key store file you will use the PKCS12 tool. You can locate the necessary files in the following locations:

- For Windows installations: <INSTALLDIR>\SAP BusinessObjects Enterprise XI 4.0\java\lib.

- For Unix installations: <INSTALLDIR>/sap_bobj/enterprise_xi40/java/lib.

1. Open a command prompt with an administrative account on your SAP BusinessObjects BI platform system.

2. Navigate to the above listed location for the PKCS12 tool.

3. You can execute the tool with the parameters shown in Table 2.7.

Parameter	Default Value
keystore	Keystore.p12
alias	Myalias
storepass	123456
dname	CN=CA
validity	365
cert	Cert.der

Table 2.7 Parameter values for PKCS12

4. Run the following command:

   ```
   java -jar PKCS12Tool.jar -keystore BOEServer.
   p12 -alias BOEServer -storepass 1111 -dname
   CN=BOESERVER,OU=PM,O=SAP,C=CA -validity 365
   ```

5. The .p12 file is being generated at the location where you started the tool.

6. You now need to export the .p12 file to a certificate. The keytool is located in the folder:

   ```
   <INSTALLDIR>/SAP BusinessObjects Enterprise XI 4.0\win64_
   x64\sapjvm\bin
   ```

7. Open a command prompt with an administrative user.

8. Run the following command:

   ```
   "<INSTALLDIR>\SAP BusinessObjects Enterprise XI 4.0\win64_
   x64\sapjvm\bin\keytool" -exportcert -keystore BOEServer.p12
   -storetype pkcs12 -file BOEServer.cert -alias BOEServer
   ```

   You need to replace the placeholder <INSTALLDIR> with the path to your installation folder. The command above shows the command used with the values from our previous steps.

9. Enter the password you previously configured via the parameter storepass; in our example, 123456.

10. The certificate is stored in the folder.

So far you created a key store file and a certificate for your SAP BusinessObjects BI platform system. In the next step you will import the certificate to your SAP system.

Importing the Certificate to the SAP System

In the next steps you will import the certificate from the SAP BusinessObjects BI platform system to the SAP system.

1. Log on to your SAP system using SAP GUI.

2. Start transaction code STRUSTSSO2 (see Figure 2.57).

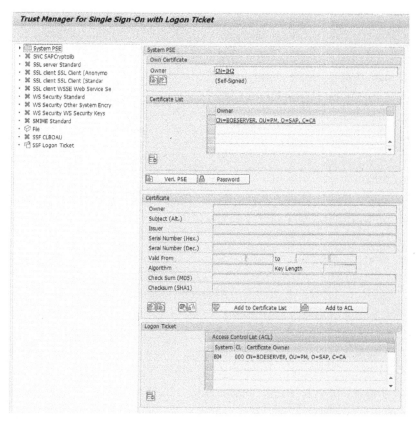

Figure 2.57 Trust Manager

3. Open the folder SYSTEM PSE.

4. Double-click the entry for your server.

5. Select the menu Certificate > Import (see Figure 2.58).

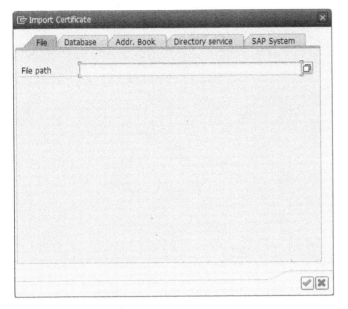

Figure 2.58 Import Certificate

6. Select the certificate file that you created previously.

7. Use the ☑ icon to confirm the settings and import the certificate (see Figure 2.59).

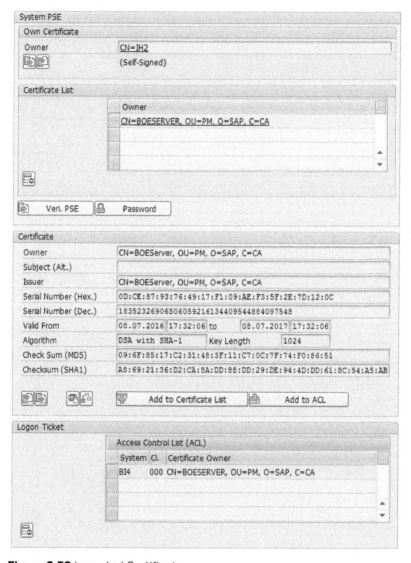

Figure 2.59 Imported Certificate

8. Click Add to Certificate List.

9. Click Add to ACL.

10. Enter a System ID for your SAP BusinessObjects BI platform system. This System ID will also be used later on in the CMC.

11. For the Client field enter 000.

12. Save your settings.

You have now imported the certificate from your SAP BusinessObjects BI platform server to your SAP system. In the next step you will configure the necessary settings in the CMC of your SAP BusinessObjects BI platform.

Setting Up SSO in the CMC
We generated the certificate and imported it into the SAP system, and we will now set up the necessary steps in the CMC.

1. Log on to the CMC with an administrative account.

2. Navigate to the area Authentication.

3. Double-click the entry SAP.

CHAPTER 2

4. Navigate to the Options tab (see Figure 2.60).

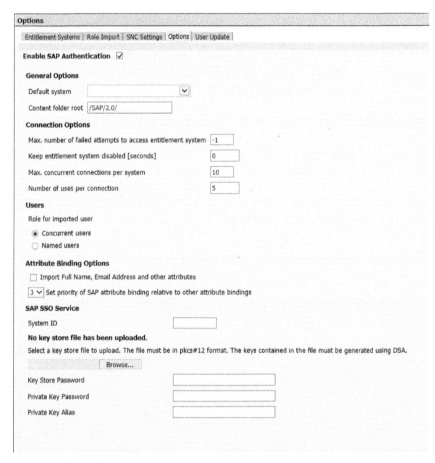

Figure 2.60 SAP Authentication options

5. Click Browse.

6. Select your key store file; in our example, BOEServer.p12.

7. Enter the configured Key Store Password. Here you enter the password you configured in the previous section.

8. Enter the configured Private Key Password. Here you enter the password you configured in the previous section.

9. Enter the configured alias for your SAP BusinessObjects BI platform system into the Private Key Alias field—in our example, BOEServer.

10. Enter the System ID you configured as part of the Trust Manager for your SAP BusinessObjects BI platform into the field System ID—in our example, BI4.

11. Click Update.

After you configured the details in the CMC, you can add the service to your SAP BusinessObjects BI platform server.

Adding SSO Token Service to the SAP BusinessObjects Enterprise System
In this section you will now add and configure the necessary services to your SAP BusinessObjects BI platform environment.

1. Log on to the CMC with an administrative account.

2. Navigate to the area Servers.

3. Select the entry Core Services.

4. Select the entry Adaptive Processing Server.

5. Select the menu Actions > Stop Server.

6. Select the entry Adaptive Processing Server.

7. Select the menu Actions > Select Services.

8. Ensure that the Security Token Service is listed in the list of Services for the Adaptive Processing Server on the right side (see Figure 2.61).

Figure 2.61 Select Services

9. Click OK.

10. Select the entry Adaptive Processing Server.

11. Select the menu Actions > Start Server.

You have now configured the Adaptive Processing Server to leverage your configuration and offered the SSO Token Service as part of your SAP BusinessObjects BI platform. At this point you have configured the SAP Authentication and you are able to import your SAP users and roles as part of your SAP BusinessObjects BI system. In the next section we will configure the SAP HANA Authentication.

2.9 Configuring SAP HANA Authentication

Starting with SAP BusinessObjects BI release 4.1 you can also set up SSO towards your SAP HANA database. In the following steps we will show how this can be achieved.

Technical Prerequisites

- Before you configure SSO with SAP HANA with SAML, you must configure the Secure Sockets Layer (SSL) for the SAP HANA system.

To set up SSO with SAP HANA, follow these steps:

1. Log on to the CMC with an administrative account.

2. Navigate to the area Applications.

3. Select the item HANA Authentication.

4. Use a right-click and select the menu Configure HANA Connections.

5. Select the menu Manage > New Connection (see Figure 2.62).

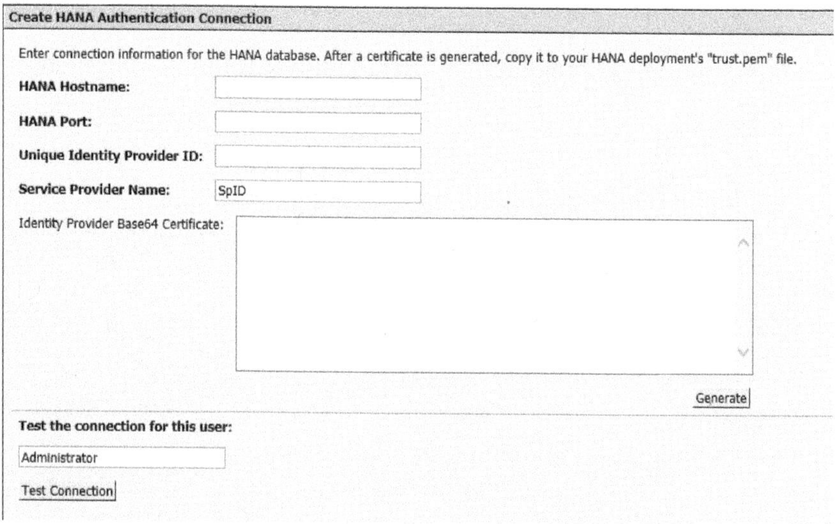

Create HANA Authentication Connection

Enter connection information for the HANA database. After a certificate is generated, copy it to your HANA deployment's "trust.pem" file.

HANA Hostname:

HANA Port:

Unique Identity Provider ID:

Service Provider Name: SpID

Identity Provider Base64 Certificate:

Generate

Test the connection for this user:

Administrator

Test Connection

Figure 2.62 SAP HANA connection

6. Enter the HANA Hostname and HANA Port according to your SAP HANA system.

7. Enter a new Unique Identity Provider ID. This ID is being used to identify your SAP BusinessObjects BI platform system.

8. Click Generate. You will notice that a certificate is being generated and shown in the Identity Provider Base64 Certificate box.

9. Click OK to save the settings.

10. Now start SAP HANA studio and log on to your SAP HANA system.

11. In SAP HANA studio select the system entry and select the menu Open SQL Console.

12. Execute the following SQL command:

```
CREATE SAML PROVIDER <%PROVIDER NAME%>
WITH
SUBJECT 'C=CA, ST=BC, O=SAP, OU=BOE, CN=<%UNIQUE
IDENTITY PROVIDER ID%> '
ISSUER 'C=CA, ST=BC, O=SAP, OU=BOE, CN=<%UNIQUE
IDENTITY PROVIDER ID%> '
```

Replace the placeholder <%UNIQUE IDENTITY PROVIDER ID%> with the value you previously defined in the CMC.

Replace the placeholder <%PROVIDER NAME%> with a SAML Provider Name you define.

13. As the next step you will have to copy the previously generated certificate from the CMC and append the content to the trust.pem file on your SAP HANA system. By default the file would be located in /usr/sap/<HANA Instance Name>/home/.ssl.

14. After you change the file, restart your SAP HANA system.

15. After the restart, log on to your SAP HANA system in SAP HANA studio.

16. In SAP HANA studio select the system entry and select the menu Open SQL Console.

17. Execute the following SQL command:

```
CREATE USER <%USERNAME%> PASSWORD
<%PASSWORD%>;
```

Replace %USERNAME% with a username of your choice and include a password for the placeholder %PASSWORD%.

18. Execute the following SQL command to enable SAML for the user:

```
ALTER USER <%USERNAME%> ENABLE SAML;
```

19. Execute the following SQL command to enable SAML for the user:

```
ALTER USER <%USERNAME%> ADD IDENTITY '<%BI4
USERNAME%>' FOR SAML PROVIDER <%PROVIDER NAME%>
```

Placeholder <%BI4 USERNAME%> needs to be replaced with the username from your SAP BusinessObjects BI system and placeholder <%PROVIDER NAME%> needs to match the provider name you entered in step 12.

Now the configuration steps on the SAP HANA side are finished and you can navigate to the CMC to validate the configuration.

1. Log on to the CMC with an administrative account.

2. Navigate to the area Applications.

3. Select the item HANA Authentication.

4. Use a right-click and select the menu Configure HANA Connections.

5. Select the previously configured SAP HANA Connection (see Figure 2.63).

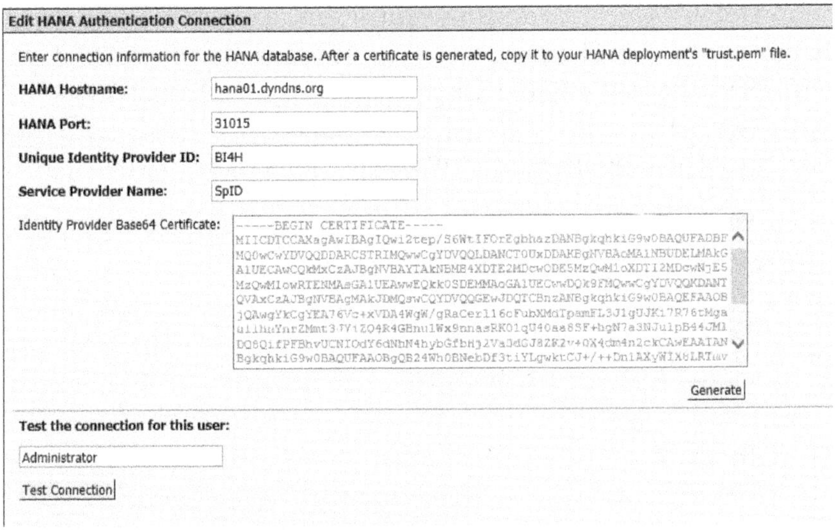

Edit HANA Authentication Connection

Enter connection information for the HANA database. After a certificate is generated, copy it to your HANA deployment's "trust.pem" file.

HANA Hostname: hana01.dyndns.org

HANA Port: 31015

Unique Identity Provider ID: BI4H

Service Provider Name: SpID

Identity Provider Base64 Certificate:

-----BEGIN CERTIFICATE-----
MIICDTCCAXagAwIBAgIQwi2tep/S6WtIFOrZgbhazDANBgkqhkiG9w0BAQUFADBF
MQ0wCwYDVQQDDARCSTRIMQswCgYDVQQLDANCT0UxDDAKEgNVBAcMA1NBUDELMAkG
A1UECAwCQQkMxCzAJBgNVBAYTAkNEME4XDTE2MDcwODE5MzQwM1oXDTI2MDcwNjE5
MzQwM1owRTENMAsGA1UEAwwEQkk0SDEMMAoGA1UECswDQk9FMQwwCgYDVQQKDANT
QVAxCzAJBgNVBAgMAkJDMQswCQYDVQQGEwJDQTCBnzANBgkqhkiG9w0BAQEFAAOB
jQAwgYkCgYEA76Vc+xVDA4WgW/gRaCer116cFubXMdIpsmFL3J1gUJKi7R76tMga
uiihuYnrZMmt3-YliZO4R4GBnuiWx9nnasRKO1qU40az85F+hgN7a3NJulpB44JM1
DQ8QifPFBhvUCNIOdY6dNhN4hybGfbHj2Va3dGJ8ZR2v+0X4dm4n2ckCAwEAATAN
BgkqhkiG9w0BAQUFAAOBgQB24Wh0BNebDf3tiYLgwktCJ+/++Dn1AXyWIXtLRTav
-----END CERTIFICATE-----

Generate

Test the connection for this user:

Administrator

Test Connection

Figure 2.63 SAP HANA connection

6. You can now enter an SAP BusinessObjects BI platform user under the Test the connection for this user: option. Click Test Connection to see if your configuration is successful.

In this section we configured the SSO with your SAP HANA system. In the next section we will set up our first set of connections towards SAP BW and SAP HANA, so that we can use these connections with the different BI client tools.

2.10 Setting Up Your First Connections

Before we can start using our BI client tools in combination with SAP BW and SAP HANA we need to set up the connections towards our systems. In this section we will configure the connection as part of our SAP BusinessObjects BI platform. In regards to the SAP HANA connection, this section will focus on using the option to establish a HTTP-based connection instead of using an ODBC/JDBC connection, because in that way we do not need to configure ODBC and JDBC details.

Setting Up an OLAP Connection for SAP BW

You can set up an OLAP connection towards SAP BW pointing to the BW system, or pointing to a particular InfoProvider, or pointing to a single Business Explorer (BEx) Query. The recommendation is to establish a connection towards the SAP BW system and provide the user the flexibility to choose the InfoProvider and query in the BI client product.

To set up an OLAP connection towards SAP BW, please follow these steps:

1. To start the configuration of the data connection, log on to the CMC.

2. Log on with an administrative account.

3. Navigate to the OLAP Connections area (see Figure 2.64).

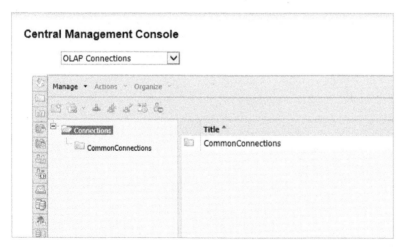

Figure 2.64 OLAP Connections

4. Use the New icon ( ) to create a new OLAP connection (see Figure 2.65).

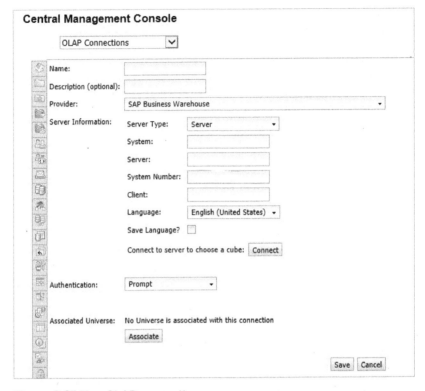

Figure 2.65 New OLAP connection

5. Enter the details for your OLAP connection according to Table 2.8 with the values for your SAP system.

Field Name	Value Description
Name	The name of the connection the user sees in the BI tool.
Description (optional)	Description for the connection. The value is optional.
Provider	The type of connection. For our example connecting to SAP BW, select the SAP Business Warehouse provider.
Server Type	Here you can select between an application server (Server) or a message server with a logon group (Group).
System	The three-digit system ID of your SAP BW system. Example: IH1
Server	This is either the full qualified name of your application server or your message server, depending on the Server Type you selected. Example: ihilgefort.dyndns.org
System Number	The two-digit system number of your SAP BW system. Example: 00
Client	The client number to which you would like to connect from your SAP BW system. Example: 800
Language	The two-letter code for the language. Example: EN for English.
Save Language	You can use this option to save the language, so that the setting in the user profile does not overwrite the setting in the connection.
Authentication	Here you can select one of three options: Prompt, SSO, or User specified.

Table 2.8 Connection details

6. Click Connect.

7. You are asked to enter your SAP credentials to log on to the SAP BW system. After doing so, you are presented with a list of InfoProviders and BEx queries. You can choose an InfoProvider or a BEx Query, or you can also leave the selection empty and have the connection point towards the SAP BW system.

8. Choose a InfoProvider or BEx Query, or click Cancel to leave the selection empty.

CHAPTER 2

9. Set the Authentication option to the value SSO.

10. Click Save.

You have now created an OLAP connection as part of your SAP BusinessObjects BI platform system, and you can now use the connection with the BI clients.

OLAP Connections

If you are creating an OLAP connection pointing to a single BEx Query, the report designer will only have access to the single BEx Query.

If you are creating an OLAP connection pointing to the SAP BW System or to an InfoProvider, the report designer can then select from a list of available BEx Queries for the SAP BW system or for the InfoProvider.

Setting Up an OLAP Connection for SAP HANA

In the following steps we will configure our connection towards SAP HANA as an OLAP connection on our SAP BusinessObjects BI platform.

To set up a OLAP connection towards SAP HANA, follow these steps:

1. To start the configuration of the data connection, log on to the CMC.

2. Log on with an administrative account.

3. Navigate to the OLAP Connections area.

4. Use the New icon (⊕) to create a new OLAP connection.

5. Set the Provider to the option SAP HANA http (see Figure 2.66).

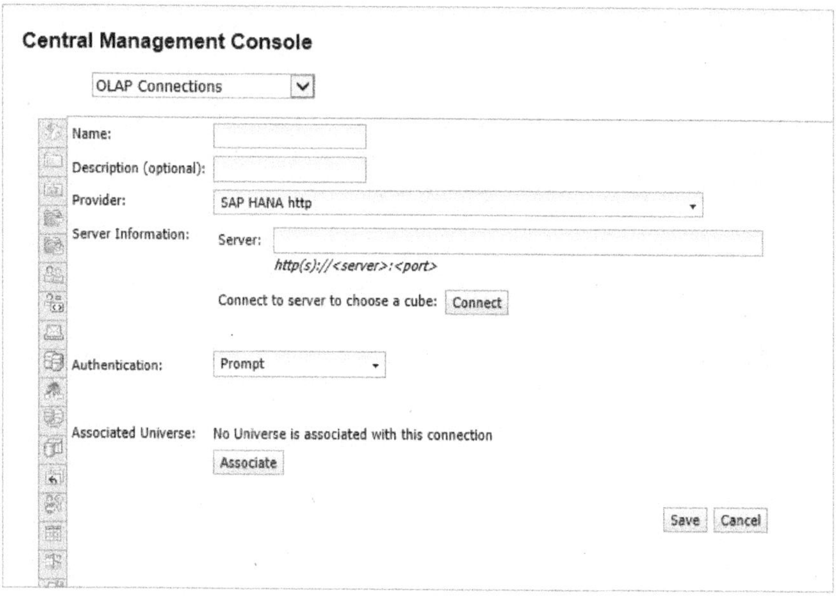

Figure 2.66 New SAP HANA HTTP connection

6. Enter the details for your OLAP connection according to Table 2.9 with the values for your SAP HANA system.

Field Name	Value Description
Name	The name of the connection the user sees in the BI tool.
Description (optional)	Description for the connection. The value is optional.
Provider	The type of connection. For our example connecting to SAP HANA, select the SAP HANA http option.
Server	Here you need to enter the URL to your SAP HANA system in the format: http(s)://<HANA Server Name>:<port> The default port would be 80<Instance Number>, so for an instance number 00 the port would be 8000.
Authentication	Here you can select one of three options: Prompt, SSO, or User specified.

Table 2.9 Connection details

7. Click Connect.

8. You are asked to enter your SAP credentials to log on to the SAP HANA system. After doing so, you are presented with a list of SAP HANA models and you can choose a specific model, or you can also leave the selection empty and have the connection point towards the SAP HANA system.

9. Choose an SAP HANA model, or click Cancel to leave the selection empty.

10. Set the Authentication option to the value SSO.

11. Click Save.

You have now created an OLAP connection pointing to SAP HANA as part of your SAP BusinessObjects BI platform system, and you can now use the connection with the BI clients.

2.11 Summary

In this chapter we reviewed the installation and configuration of the SAP BusinessObjects BI platform as well as the BI client tools. In addition, we configured the SAP Authentication and established connections towards SAP BW and SAP HANA. In the next chapter we will take a closer look at Analysis Office.

SAP BusinessObjects Analysis, edition for Microsoft Office (Analysis Office)

In this chapter you will learn about the data-connectivity options and the level of support for the metadata in SAP BW and SAP HANA, and you will create your very first workbook using SAP BusinessObjects Analysis, edition for Microsoft Office (Analysis Office).

3.1 Data-Connectivity Overview

Figure 3.1 shows the data connectivity between Analysis Office and your SAP BW, SAP ERP, and SAP HANA systems.

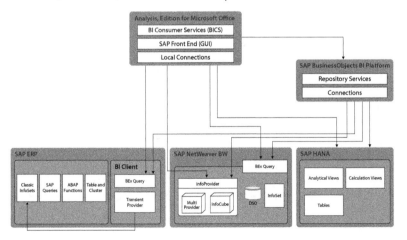

Figure 3.1 Data connectivity

Analysis Office can be leveraged in combination with the SAP BusinessObjects BI platform and in a so-called "lean deployment," where Analysis Office is deployed without the use of the SAP BusinessObjects BI platform, or in combination with SAP BW as the platform (not to be confused with SAP BW as the data source). In regards to data connectivity, Analysis Office is able to:

- Leverage a direct access method using BI Consumer Services (BICS) to the InfoProvider and BEx Queries from your SAP BW system.

- Leverage a direct access method using BICS with BEx Queries based on transient providers in the SAP ERP system. The transient provider requires SAP ERP Central Component (ECC) 6.0, enhancement package 05.

- Leverage the shared connections from the SAP BusinessObjects BI platform using the direct access method based on BICS towards the SAP BW system.

- Establish a direct link to SAP HANA using an HTTP(s) connection via the SAP BusinessObjects BI platform and as a local connection.

Analysis Office can be deployed with or without the deployment of the SAP BusinessObjects BI platform and with or without the usage of SAP BW as the platform. Below we list some of the disadvantages of deploying Analysis Office without the SAP BusinessObjects BI platform and some of the advantages that the SAP BusinessObjects BI platform can add to your overall BI landscape.

The disadvantages for a deployment without the SAP BusinessObjects BI platform are:

- All connections towards any SAP BW system have to be created on each client computer as part of the SAP front end.

- Microsoft Excel and PowerPoint documents can be shared only on some form of the central shared folder, as there is no central repository.

- The SAP front end has to be deployed on each client computer that will leverage Analysis Office.

- Such a deployment does not offer any form of administration services in terms of access to Microsoft Excel and PowerPoint documents.

- Such a deployment does not leverage any form of a central lifecycle mechanism, which means that there is no automated way of populating shared documents from development to a quality assurance or production environment.

- A deployment without the SAP BusinessObjects BI platform does not provide any integrations with other BI clients, such as SAP BusinessObjects Design Studio or SAP BusinessObjects Lumira.

- A deployment without the SAP BusinessObjects BI platform does not provide any integration with the SAP Mobile BI solution, to provide your users with a mobile solution for their workbooks.

The advantages of an integrated deployment with the SAP BusinessObjects BI platform are:

- Users can centrally save, open, and share documents created with Analysis Office by storing them via the SAP BusinessObjects BI platform system.

- Administrators can create connections to the SAP BW system centrally and control access to those connections in the SAP BusinessObjects BI platform system.

- Administrators can assign authorizations to centrally stored documents created with Analysis Office and control access to these documents on a user and group level.

- Administrators can leverage the SAP BusinessObjects Lifecycle management console to follow documents and connections from development, to quality assurance, to production.

- When Analysis Office is deployed combined with the SAP BusinessObjects BI platform, there is no need to have the SAP front end deployed on the client, because users can share connections via the SAP BusinessObjects BI platform system.

- Users can leverage the integration of Analysis Office with the other BI clients, such as SAP BusinessObjects Analysis, edition for

CHAPTER 3

Online Analytical Processing (OLAP), and SAP BusinessObjects Design Studio, and share their content across the BI clients.

Analysis Office can also be deployed in combination with SAP BW as the platform without the use of the SAP BusinessObjects BI platform. In such a scenario the user can store the workbooks into the SAP BW repository and the necessary security is handled by the authorizations in SAP BW. In a scenario in which you deploy Analysis Office in combination with SAP BW without the use of the SAP BusinessObjects BI platform:

- You would manage the data connectivity using the SAP front end on each desktop system that has Analysis Office installed.

- You would manage the lifecycle management of your workbooks using SAP transports.

In this section we reviewed the different data connectivity options available with Analysis Office and we looked at the different options for deploying Analysis Office. In the next section you will learn about the level of support of your existing metadata from SAP BW in combination with Analysis Office.

3.2 Supported and Unsupported SAP BW Elements

In this section we review the level of support for your existing metadata inside the SAP BW system for Analysis Office. Table 3.1 shows which of the objects are supported when using Analysis Office as the BI client tool.

SAP BW Metadata	Direct Access Using BICS
Direct access to InfoCubes and MultiProviders	Yes
Access to BEx Queries	Yes
Characteristic values	
Key	Yes
Short description	Yes
Medium and long descriptions	Yes

Table 3.1 Supported and unsupported BEx Query features for Analysis Office *(continues)*

SAP BW Metadata	Direct Access Using BICS
BEx Query Features	
Support for hierarchies	Yes
Support for free characteristics	Yes
Support for calculated and restricted key figures	Yes
Support for currencies and units	Yes
Support for custom structures	Yes
Support for formulas and selections	Yes
Support for filter	Yes
Support for display and navigational attributes	Yes
Support for conditions in rows	Yes
Support for conditions in columns	Yes
Support for conditions for fixed characteristics	Yes
Support for exceptions	Yes
Compounded characteristics	Yes
Constant selection	Yes
Default values in BEx Query	Yes
Number scaling factor	Yes
Number of decimals	Yes
Calculate rows as (local calculation)	Yes
Sorting	Yes
Hide/Unhide	Yes
Display as hierarchy	Yes
Reverse sign	Yes
Support for reading master data	Yes
Data Types	
Support for CHAR (characteristics)	Yes
Support for NUMC (characteristics)	Yes
Support for DATS (characteristics)	Yes
Support for TIMS (characteristics)	Yes
Support for numeric key figures such as Amount and Quantity	Yes
Support for Date (key figures)	Yes

Table 3.1 Supported and unsupported BEx Query features for Analysis Office *(continues)*

CHAPTER 3

SAP BW Metadata	Direct Access Using BICS
Support for Time (key figures)	Yes
BEx Query Variables—Processing Type	
User input	Yes
Authorization	Yes
Replacement path	Yes
SAP exit/custom exit	Yes
Precalculated value set	Yes
General Features for Variables	
Support for optional and mandatory variables	Yes
Support for key date dependencies	Yes
Support for default values	Yes
Support for variable variants	Yes
Support for personalized values	No
BEx Query Variables—Variable Type	
Single value	Yes
Multi-single value	Yes
Interval value	Yes
Selection option	Yes
Hierarchy variable	Yes
Hierarchy node variable	Yes
Hierarchy version variable	Yes
Text variable	Yes
EXIT variable	Yes
Single key date variable	Yes
Multiple key dates	Yes
Formula variable	Yes

Table 3.1 Supported and unsupported BEx Query features for Analysis Office (continued)

In Table 3.2 you can see how the direct-access method with the BICS option uses elements from the BEx Query and how the objects are mapped to the navigation panel for Analysis Office.

BEx Query Element	Analysis Office
Characteristic	For each characteristic you'll receive a field, and with the menu members you can decide which part of the characteristic is shown as part of the overall result.
Hierarchy	Each available hierarchy is shown as an external hierarchy and can be leveraged as part of the crosstab. In addition, you can leverage hierarchy levels as part of your crosstab; for example, you can show all members of Level 2 of the hierarchy.
Key figure	Each key figure is shown with the unit and scaling factor information.
Calculated/Restricted key figure	Each calculated and restricted key figure is treated like a key figure. The user does not have access to the underlying definition in Analysis Office.
Filter	Filters are applied to the underlying query and are visible in the navigation panel as part of the Filter area in the Information tab.
Display attribute	Display attributes become standard fields in the navigation panel and are grouped as subordinates of the linked characteristic.
Navigational attribute	Navigational attributes are treated the same way as characteristics.
Variable	Each variable with the property Ready for Input results in a prompt. You can leverage the Prompts menu to provide the necessary input values.
Custom structure	A custom structure is available as an element in the navigation panel and each structure element can be selected or de-selected for the report.

Table 3.2 BEx Query metadata mapping for Analysis Office

CHAPTER 3

In addition to the details on how Analysis Office supports your existing metadata from SAP BW, we would like to highlight the fact that the terms that are used in Analysis Office are different from those that might be used in the BEx Analyzer. Table 3.3 maps those terms.

BEx Query Terms	Analysis Office
Key figure	Measure
Characteristic	Dimension
Variable	Prompt
Characteristic value	Member
Condition	Filter by measure
Exception	Conditional formatting

Table 3.3 Terms in Analysis Office

As you can see in Table 3.1 and Table 3.2, Analysis Office strongly supports existing metadata from SAP BW and provides you with a feature-rich premium successor to your BEx Analyzer. In the next section, you will see how Analysis Office is able to support your SAP HANA metadata.

3.3 Supported and Unsupported SAP HANA Elements

In this section we review the level of support for your existing metadata from SAP HANA when using Analysis Office. Table 3.4 shows which of the objects are supported when using Analysis Office as the BI client tool.

SAP HANA Metadata	Direct Access to SAP HANA
Analytical Model	Yes
Calculation Model	Yes
Attribute Model	Yes
Dimension Key	Yes
Dimension description (label column)	Yes
Unit/Currency	Yes
Calculated column	Yes
Restricted column	Yes
Variable	Yes
Input Parameter	Yes
Level-Based Hierarchy	Yes
Parent-Child Hierarchy	Yes

Table 3.4 Supported elements from SAP HANA

After reviewing the support for your SAP HANA-based metadata, we will learn how to set up the local connections in Analysis Office.

CHAPTER 3

3.4 Setting Up Local Connections

In addition to using the shared OLAP connections from your SAP BusinessObjects BI platform, Analysis Office is also capable of using locally defined connections.

Setting up the local connection for SAP BW is as simple as adding your SAP entry to the SAP logon from your SAP GUI. Analysis Office will then retrieve the list of system entries and make them available, so that you can establish a local connection.

When it comes to setting up the local HTTP(s) connection to your SAP HANA there are a few more steps involved and there are a few technical prerequisites:

- You need to use SAP HANA Support Package 09 or higher.

- The SAP HANA Info Access Service with delivery unit HCO_INA_ SERVICE is deployed.

- The SAP HANA Info Access Service can be accessed with authentication method Basic (see SAP Note 2193057 for details).

- The role sap.bc.ina.service.v2.userRole::INA_USER is assigned to your users (see SAP Note 2097965 for details).

To set up a local SAP HANA HTTP connection, follow these steps:

1. Start SAP BusinessObjects Analysis for Microsoft Excel.

2. Navigate to the Analysis ribbon (see Figure 3.2).

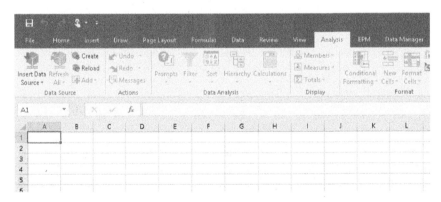

Figure 3.2 Analysis ribbon

3. Select the menu Insert Data Source > Select Data Source.

4. If you are prompted to log on to the SAP BusinessObjects BI platform, click Skip.

5. You are presented with the list of connections (see Figure 3.3).

6. Use a right-click and select the menu Create New SAP HANA Connection...

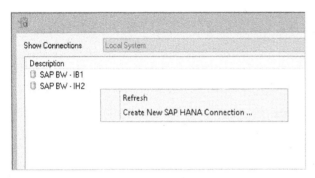

Figure 3.3 Local connections

7. You can now define the details for your SAP HANA system (see Figure 3.4).

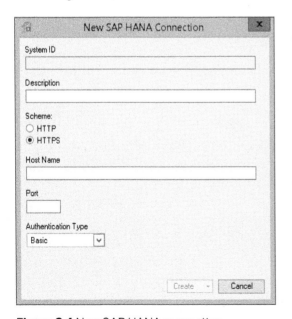

Figure 3.4 New SAP HANA connection

CHAPTER 3

141

8. You can enter the details according to Table 3.5.

Field Name	Value Description
Description	The name of the connection the user sees in the BI tool.
Scheme	Either HTTP or HTTPS.
Host Name	The fully qualified hostname of your SAP HANA system.
Port	The default port would be 80<Instance Number>, so for instance number 00 the port would be 8000.
Authentication	Here you can select one of three options: Basic, Client Certificates, or Kerberos.

Table 3.5 Connection details

9. Enter the required details and click Create.

10. The new connection is being added to the list of local connections and you can now log on to the new connection.

We are now able to establish server-side connections as well as local connections. In the next section we will create our first workbook using Analysis Office.

3.5 Creating Your First Workbook in Analysis Office

In the following steps you will learn how to use some of the basic functionality of Analysis Office. You will leverage the connection created previously in the Central Management Console (CMC) of your SAP BusinessObjects BI platform system.

For our example steps we are using a BEx Query based on the SAP NetWeaver Demo model (http://scn.sap.com/docs/DOC-8941) with the following details:

Rows:

- Product Group (0D_NW_PROD__0D_NW_PRDGP)

Columns:

- Net Value (0D_NW_NETV)

- Costs (0D_NW_COSTV)

Free Characteristics:

- Country (0D_NW_CNTRY)
- Region (0D_NW_REGIO)
- Sold-to-Party (0D_NW_SOLD)
- Product Category (0D_NW_PROD__0D_NW_PRDCT)
- Product (0D_NW_PROD)
- Cal. Year / Month (0CALMONTH)

1. Start SAP BusinessObjects Analysis for Microsoft Excel.

2. Navigate to the Analysis tab.

3. Select the menu Insert Data Source > Select Data Source and you see Figure 3.5.

Figure 3.5 Log on to SAP BusinessObjects BI platform

4. You are asked to log on to your SAP BusinessObjects BI platform system or you can use the Skip option and leverage the connection with your local SAP GUI. In our example we use the shared connection from our SAP BusinessObjects BI platform system.

Platform Preferences

In Analysis Office you can leverage the menu path File > Settings > Platform to configure if you would like to mainly work with the SAP BusinessObjects BI platform or with SAP BW as the platform or if you would prefer to leave it as a selectable option.

5. In the field for the Web Service URL replace the placeholders with the values for your SAP BusinessObjects system.

   - BOE SERVER HOSTNAME: Here you need to enter the name of your Java application server that is used as part of your SAP BusinessObjects BI platform system.
   - PORT: Here you need to enter the port of your Java application server that is used as part of your SAP BusinessObjects BI platform system.

6. Click Options.

7. Enter the name of your system into the SYSTEM field. Here you should enter the name of your CMS.

8. Select SAP as the AUTHENTICATION.

9. Enter your SAP credentials. Because you cannot enter the SAP system ID and the client number in separate fields, you need to enter your SAP user in the following syntax: <SAP System ID>~<SAP Client Number>/<SAP User>.

10. Enter your password.

11. Click OK. The list of available connections from the SAP BusinessObjects BI platform system is shown (see Figure 3.6).

Figure 3.6 Connections

12. Select the previously established connection.

13. Click OK. You are asked to log on to the SAP system again.

14. Enter the SAP credentials.

15. Click OK.

16. You can now search for your BEx Query in the different tabs (see Figure 3.7). Select your BEx Query and click OK.

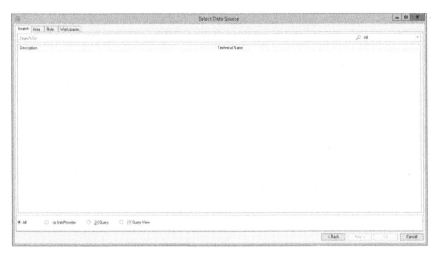

Figure 3.7 Select the data source

17. You are presented with the data according to the layout defined by the BEx Query (see Figure 3.8).

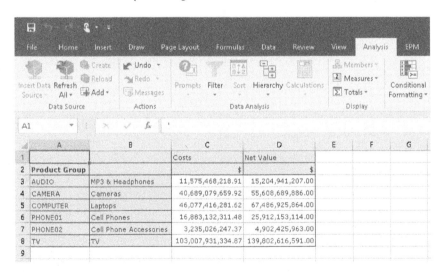

Figure 3.8 Analysis workbook

18. Use a right-click on the Measures in the Columns in the Navigation Panel (see Figure 3.9).

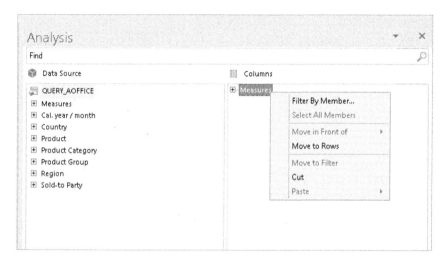

Figure 3.9 Measures

19. Select the menu Filter by Member... (see Figure 3.10).

Figure 3.10 Filter by Member

20. You are presented with the list of available members. Select only the entry Net Value and de-select all other entries.

21. Click OK.

22. Now navigate to the Analysis ribbon and click Pause Refresh (see Figure 3.11).

Figure 3.11 Pause Refresh

23. Drag and drop the dimension Product from the navigation panel to the Rows section, so that the Rows display Product Group and Product. You will notice that your crosstab is not being updated, because we activated the Pause Refresh option.

CHAPTER 3

24. Now click again on the Pause Refresh item in the Analysis ribbon and your crosstab is updated (see Figure 3.12).

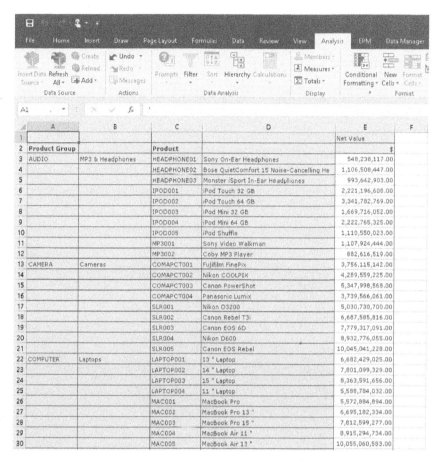

Figure 3.12 Analysis workbook

25. The Pause Refresh function allows you to defer the update of your workbook until you enable it. In this way, you can perform several navigation steps and then perform a single update to reflect all changes.

26. Select a member of dimension Product as part of your crosstab.

27. Select the menu Members from the Analysis ribbon (see Figure 3.13).

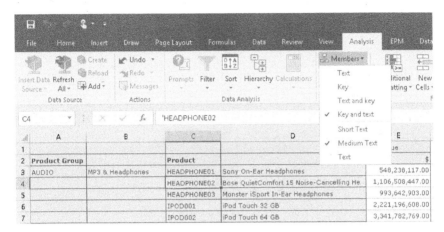

Figure 3.13 Menu Members

28. Select the Text entry and choose Text in the lower part as well.

29. Select an entry from measure Net Value.

30. Select the Measures menu from the Analysis ribbon (see Figure 3.14).

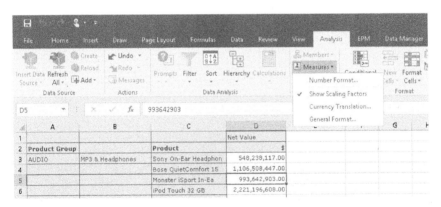

Figure 3.14 Menu Measures

31. Select the menu Number Format...

32. Format your measure with a scaling factor of 1000 and 0 decimal places (see Figure 3.15).

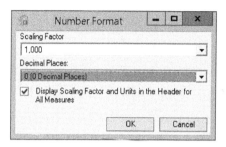

Figure 3.15 Number Format

33. Click OK.

34. Select the column header for the Net Value measure.

35. Select the menu Filter > Filter by Measure > Most Detailed Dimension in Rows > Edit (see Figure 3.16).

Figure 3.16 Filter by Measure

36. Select the TOP N option.

37. Enter the value 5.

CHAPTER 3

38. Click Add. You are presented with the top five materials (see Figure 3.17).

	A	B	C	D
1				Net Value
2	**Product Group**		**Product**	* 1,000 $
3	**Overall Result**			**308,917,753**
4	AUDIO	MP3 & Headphones	Result	15,204,941
5			iPod Touch 32 GB	2,221,197
6			iPod Touch 64 GB	3,341,783
7			iPod Mini 32 GB	1,669,716
8			iPod Mini 64 GB	2,222,765
9			iPod Shuffle	1,110,550
10	CAMERA	Cameras	Result	55,608,690
11			Canon PowerShot	5,347,999
12			Canon Rebel T3i	6,687,586
13			Canon EOS 6D	7,779,317
14			Nikon D600	8,932,776
15			Canon EOS Rebel	10,045,041
16	COMPUTER	Laptops	Result	67,486,926
17			14 " Laptop	7,801,099
18			15 " Laptop	8,363,592
19			MacBook Pro 15 "	7,812,599
20			MacBook Air 11 "	8,915,295
21			MacBook Air 13 "	10,055,061
22	PHONE01	Cell Phones	Result	25,912,153
23			Samsung Galaxy Ac	2,223,888
24			Samsung Galaxy S II	2,784,496
25			iPhone 5	2,781,478
26			Blackberry Q10	3,343,351
27			Blackberry Z10	3,343,579
28	PHONE02	Cell Phone Accessories	Result	4,902,426
29			Bluetooth Car Kit	660,175
30			Bose Bluetooth	884,349
31			Jawbone Bluetooth	881,637
32			Sennheiser Bluetooth	1,110,355
33			LG Case	437,538
34	TV	TV	Result	139,802,617
35			48 " LCD TV	12,307,645
36			51 " LCD TV	16,768,475
37			55 " Plasma TV	14,533,826
38			60 " Plasma TV	16,773,221
39			70 " Plasma TV	20,161,359

Figure 3.17 Top 5

39. Now use a drag-and-drop navigation and remove the dimension Product Group from the Rows area in the navigation panel.

40. Now drag and drop the Sold-to Party dimension on top of the Product dimension in the navigation panel so that it replaces the Product dimension. You are presented with the top five Sold-to Party members (see Figure 3.18).

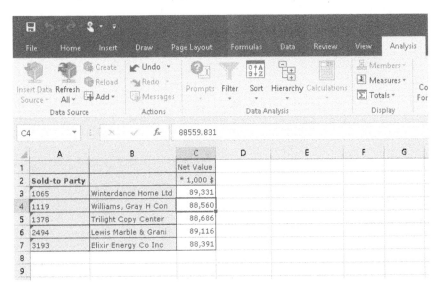

Figure 3.18 Top 5 Sold-to Parties

41. Select a value in the column Net Value.

42. Select the menu Filter > Filter by Measure > Most Detailed Dimension in Rows > Reset.

43. Select the menu Conditional Formatting > New (see Figure 3.19).

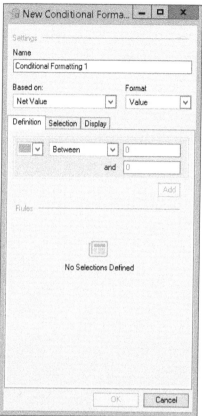

Figure 3.19 Conditional formatting

44. Enter Revenue Highlighting as the Name.

45. Set the Format to Status Symbol.

46. Add a definition for the values Greater Than 10.000.000 as Green (1). (Enter the value without separators!)

47. Click Add.

48. Add a definition for the values between 1.000.000 and 5.000.000 as Orange (4). (Enter the value without separators!)

49. Click Add (see Figure 3.20).

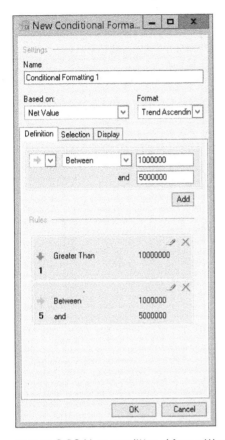

Figure 3.20 New conditional formatting

Highlighting Values

Note that depending on your data you might have to adjust the threshold values.

50. Click the Display tab.

51. Activate the options Data Cells and Row Headers.

52. Click OK.

53. Drag and drop the dimension Country to the Rows so that the Rows display the dimension Country first and then the dimension Sold-to Party.

54. Select a member from dimension Country in the crosstab.

55. Select the menu Hierarchy > Compact Display in Rows.

56. You can now open and close the levels (see Figure 3.21).

	A	B	C
1			Net Value
2	Country		* 1,000 $
3	Overall Result		⬇ 308,917,753
4	[-] AE	United Arab Emirates	⬇ 2,151,560
5	1000	Z & Z Warehouse Inc	⬇ 84,044
6	1144	Wilberforce College	⬇ 81,889
7	1288	Viking Press Co Inc	⬇ 84,548
8	1432	Telephon Enterprises	⬇ 80,814
9	1576	Sir Donough, John C	⬇ 82,682
10	1720	Second Cafe Co Inc	⬇ 81,286
11	1864	Reynolds Tools & Man	⬇ 84,502
12	2008	Poseidon Services Co	⬇ 86,162
13	2152	Newman Brothers Inc	⬇ 85,607
14	2296	Mercantil Housing In	⬇ 81,564
15	2440	Line Printing Inc	⬇ 84,540
16	2584	John's Discount Outl	⬇ 85,315
17	2728	Health Service of Co	⬇ 80,603
18	2872	Golden Oak Electroni	⬇ 81,705
19	3016	Fowler Housing Group	77,615
20	3160	Eoc Head Start	79,813
21	3304	Donaldson Store Grou	⬇ 83,564
22	3448	Coldwell College Gro	⬇ 81,874
23	3592	Carnavoran Foods Ltd	⬇ 85,730
24	3736	Bridge Clinic Co Inc	⬇ 82,711
25	3880	Blue Star University	⬇ 82,227
26	4024	Black Rock Residenti	⬇ 81,731
27	4168	Best Result Publishi	⬇ 83,521
28	4312	Barnes Industries G	⬇ 81,510
29	4456	Alside Enterprise Lt	⬇ 83,561
30	4600	Abstract Iron Group	⬇ 82,442
31	[+] AO	Angola	⬇ 2,173,539
32	[+] AR	Argentina	⬇ 2,145,655
33	[+] AT	Austria	⬇ 2,161,579
34	[+] AU	Australia	⬇ 4,290,505
35	[+] AZ	Azerbaijan	⬇ 2,164,602
36	[+] BB	Barbados	⬇ 2,159,598
37	[+] BD	Bangladesh	⬇ 2,163,588
38	[+] BE	Belgium	⬇ 2,152,315
39	[+] BG	Bulgaria	⬇ 2,144,566

Figure 3.21 Compact display

CHAPTER 3

57. Add several empty rows on the top of your spreadsheet.

58. Now open the tab Information in the navigation panel (see Figure 3.22).

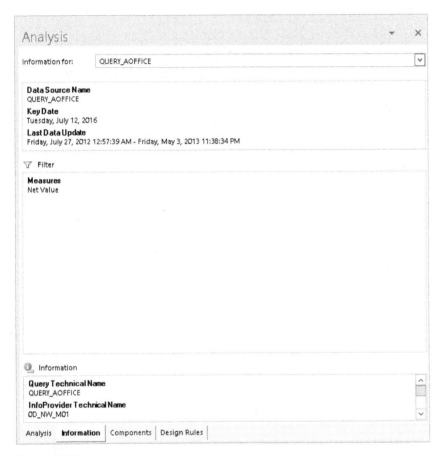

Figure 3.22 Information

59. Drag and drop all the elements from the Information Area to the top of the spreadsheet. You can simply select the item in the Information Area and drag and drop it on the spreadsheet.

60. Select a cell in the crosstab.

61. Now select the menu Smart Copy in the ribbon.

62. Start SAP BusinessObjects Analysis for Microsoft PowerPoint.

63. In Microsoft PowerPoint select the option to create a new blank presentation.

64. Choose the Blank layout for the slide.

65. Navigate to the Analysis ribbon.

66. Select the menu Smart Paste.

67. Select the option Smart Paste as Table.

68. Log on to your SAP BusinessObjects BI Server with your SAP credentials and SAP authentication.

69. You are presented with the Fit Table dialog and you can decide to either abbreviate the table or to split the table across several slides (see Figure 3.23).

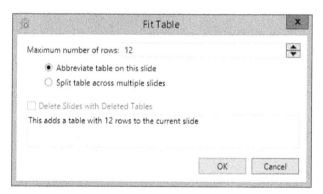

Figure 3.23 Fit Table

70. Select the option Abbreviate table on this slide.

71. Set the Maximum Number of Rows to 10.

72. Click OK.

73. You are presented with a default layout (see Figure 3.24).

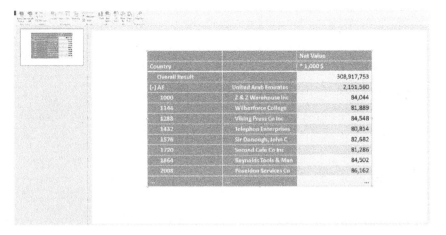

Figure 3.24 Analysis for PowerPoint

74. Select a member in the table.

75. Select the menu item Fit Table in the Analysis ribbon.

76. You can now select how your table is being treated inside Microsoft PowerPoint. You can choose to abbreviate the table or to split the table.

77. Select the option Split.

78. Set the maximum number of rows to 10.

Analysis Office tells you that you need several slides based on the amount of data.

79. Click OK.

80. Additional slides are created based on the information that you created in your crosstab. You can use all the common Microsoft PowerPoint options to format the design and layout.

81. Close Microsoft PowerPoint. (We will not save the slides.)

82. Now navigate back to SAP BusinessObjects Analysis for Microsoft Excel.

83. Open the Excel menu File (see Figure 3.25).

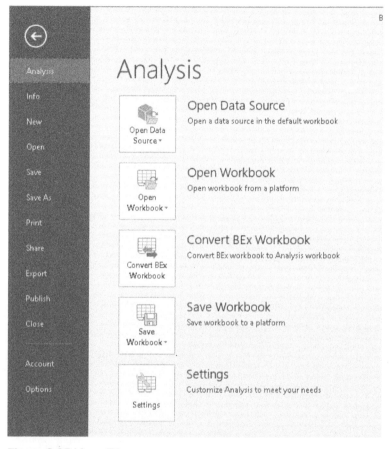

Figure 3.25 Menu File

84. Here you can save your workbook to or open it from the BI platform.

In this section you learned some of the basic steps of Analysis Office. In the next chapter we will look into more details of SAP BusinessObjects Design Studio.

3.6 Summary

In this chapter you learned how you can leverage Analysis Office in combination with data from your SAP BW system. You also learned how the product supports the existing metadata from your SAP BW system and your SAP HANA system. In the next chapter you will learn more about SAP BusinessObjects Design Studio.

CHAPTER 4

SAP BusinessObjects Design Studio (Design Studio)

In this chapter you will learn about the data-connectivity options and the level of support for the metadata in SAP BW and SAP HANA, and you will create your dashboard using Design Studio.

4.1 Data-Connectivity Overview

Figure 4.1 shows the data connectivity between Design Studio and your SAP BW, SAP ERP, and SAP HANA systems.

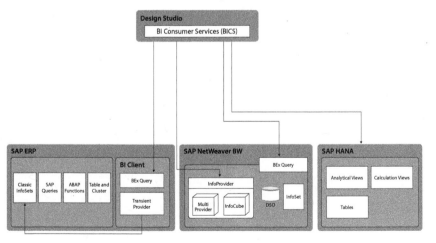

Figure 4.1 Data connectivity

Design Studio is able to leverage the direct connection towards the BEx Queries in your SAP BW system as well as use the direct connection to the BEx Query in the BI client of your SAP ERP system. In addition to the connectivity towards SAP BW, Design Studio is also able to connect directly to SAP HANA models. In regards to data connectivity, Design Studio is able to:

- Leverage a direct access method using BI Consumer Services (BICS) to the InfoProvider and BEx Queries from your SAP BW system.

- Leverage a direct access method using BICS with BEx Queries based on transient providers in the SAP ERP system. The transient provider requires SAP ERP Central Component (ECC) 6.0, enhancement package 05.

- Leverage the shared connections from the SAP BusinessObjects BI platform using the direct access method based on BICS towards the SAP BW or SAP ERP system.

- Establish a direct link to SAP HANA using an HTTP(s) connection via the SAP BusinessObjects BI platform and as a local connection. Design Studio is able to connect to the SAP HANA models.

Deployment Platform

Technically speaking, Design Studio can be deployed with SAP NetWeaver, SAP HANA, or the SAP BusinessObjects BI platform as the platform for Design Studio. Since the release 1.6 of Design Studio, SAP has announced that the only platform moving forward will be SAP BusinessObjects BI, and with release 2.0 of Design Studio SAP NetWeaver and SAP HANA will not be supported any longer as platforms.

In this section we reviewed the different data-connectivity options available with Design Studio. In the next section you will learn about the level of support of your existing metadata from SAP BW in combination with Design Studio.

4.2 Supported and Unsupported SAP BW Elements

In this section we review the level of support for your existing metadata inside the SAP BW system for Design Studio. Table 4.1 shows which of the objects are supported when using Design Studio as the BI client tool.

SAP BW Metadata	Direct Access Using BICS
Direct access to InfoCubes and MultiProviders	Yes
Access to BEx Queries	Yes
Characteristic Values	
Key	Yes
Short description	Yes
Medium and long descriptions	Yes
BEx Query Features	
Support for hierarchies	Yes
Support for free characteristics	Yes
Support for calculated and restricted key figures	Yes
Support for currencies and units	Yes
Support for custom structures	Yes
Support for formulas and selections	Yes
Support for filters	Yes
Support for display and navigational attributes	Yes
Support for conditions in rows	Yes
Support for conditions in columns	Yes
Support for conditions for fixed characteristics	Yes
Support for exceptions	Yes
Compounded characteristics	Yes
Constant selection	Yes
Default values in BEx Query	Yes
Number scaling factor	Yes
Number of decimals	Yes
Calculate rows as (local calculation)	Yes
Sorting	Yes
Hide/Unhide	Yes

Table 4.1 Supported and unsupported BEx Query features for Design Studio
(continues)

CHAPTER 4

SAP BW Metadata	Direct Access Using BICS
Display as hierarchy	Yes
Reverse sign	Yes
Support for reading master data	Yes
Data Types	
Support for CHAR (characteristics)	Yes
Support for NUMC (characteristics)	Yes
Support for DATS (characteristics)	Yes
Support for TIMS (characteristics)	Yes
Support for numeric key figures such as Amount and Quantity	Yes
Support for Date (key figures)	Yes
Support for Time (key figures)	Yes
BEx Variables—Processing Type	
User input	Yes
Authorization	Yes
Replacement path	Yes
SAP exit/custom exit	Yes
Precalculated value set	Yes
General Features for Variables	
Support for optional and mandatory variables	Yes
Support for key date dependencies	Yes
Support for default values	Yes
Support for variable variants	No
Support for personalized values	No
BEx Variables—Variable Type	
Single value	Yes
Multi-single value	Yes
Interval value	Yes
Selection option	Yes
Hierarchy variable	Yes
Hierarchy node variable	Yes
Hierarchy version variable	Yes
Text variable	Yes
EXIT variable	Yes

Table 4.1 Supported and unsupported BEx Query features for Design Studio *(continues)*

CHAPTER 4

SAP BW Metadata	Direct Access Using BICS
Single key date variable	Yes
Multiple key dates	Yes
Formula variable	Yes

Table 4.1 Supported and unsupported BEx Query features for Design Studio (continued)

In Table 4.2 you can see how the direct access method using the BICS option uses elements from the BEx Query and how the objects are mapped to Design Studio.

BEx Query Element	Design Studio
Characteristic	For each characteristic you'll receive a field, and, with the context menu in the Edit Initial View panel, you can decide which part of the characteristic is shown as part of the overall result.
Hierarchy	Each available hierarchy is shown as an external hierarchy and can be leveraged as part of the crosstab. In the Edit Initial View window, you have the option to activate hierarchies for the display as part of your dashboard.
Key figure	Each key figure is shown with the unit and scaling factor information.
Calculated/ Restricted key figure	Each calculated and restricted key figure is treated like a key figure. The user does not have access to the underlying definition in Design Studio.
Filter	Filters are applied to the underlying query and you can receive the list of configured filter values using the scripting in Design Studio.
Display attribute	Display attributes become standard fields and can be included into the display of a crosstab.
Navigational attribute	Navigational attributes are treated the same way as characteristics.
Variable	Each variable with the property Ready for Input results in a prompt. You can leverage the scripting and the prompting screen to set values for the BEx Query variables.
Custom structure	A custom structure is available as an element in the Edit Initial View panel and you can decide which elements of the structure will be used in the dashboard.

Table 4.2 BEx Query metadata mapping for Design Studio

CHAPTER 4

As you can see in Table 4.1 and Table 4.2, Design Studio strongly supports existing metadata from SAP BW and delivers on the promise to become the successor for the Web Application Designer as well as SAP Dashboards. In the next section, you will learn how Design Studio is able to support your SAP HANA metadata.

4.3 Supported and Unsupported SAP HANA Elements

In this section we review the level of support for your existing metadata from SAP HANA using Design Studio. Table 4.3 shows which objects are supported when using Design Studio as the BI client tool.

SAP HANA Metadata	Direct Access to SAP HANA
Analytical Model	Yes
Calculation Model	Yes
Attribute Model	Yes
Dimension key	Yes
Dimension description (label column)	Yes
Unit/Currency	Yes
Calculated column	Yes
Restricted column	Yes
Variable	Yes
Input Parameter	Yes
Level-Based Hierarchy	Yes
Parent-Child Hierarchy	Yes

Table 4.3 Supported elements from SAP HANA

After reviewing the support for your SAP HANA-based metadata, we will learn how to set up the local connections in Design Studio.

4.4 Setting Up Local Connections

Design Studio is able to leverage the shared connections from your SAP BusinessObjects BI platform as well as connections configured locally. Setting up the local connection for SAP BW can be done using SAP GUI or it can be done inside the Design Studio designer. When it comes to setting up the local HTTP(s) connection to your SAP HANA system, there are a few more steps involved and a few technical prerequisites:

- You need to use SAP HANA Support Package 09 or higher.

- The SAP HANA Info Access Service with delivery unit HCO_INA_ SERVICE is deployed.

- The SAP HANA Info Access Service can be accessed with the authentication method Basic (see SAP Note 2193057 for details).

- The role sap.bc.ina.service.v2.userRole::INA_USER is assigned to your users (see SAP Note 2097965 for details).

CHAPTER 4

To set up a local SAP BW connection for Design Studio, follow these steps:

1. Start the Design Studio designer.

2. Navigate to the menu Tools > Preferences (see Figure 4.2).

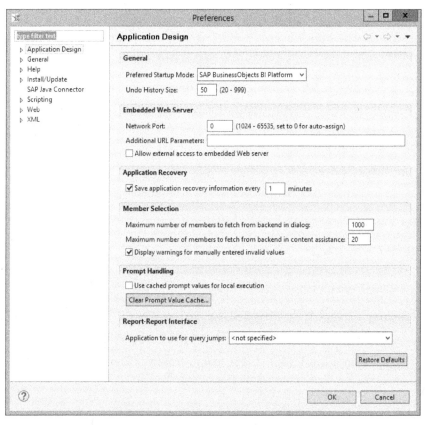

Figure 4.2 Preferences

3. Select the area Application Design.

4. Select the entry Backend Connections (see Figure 4.3).

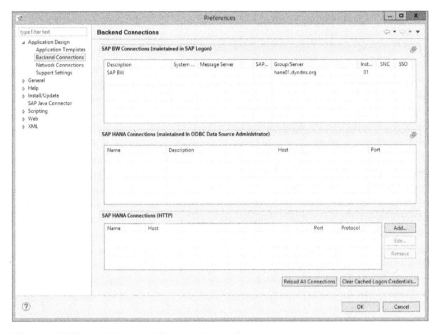

Figure 4.3 Select Backend Connections

5. In the SAP BW Connections area click the Launch SAP Logon ( ) icon (top-right corner).

6. SAP GUI will be launched and you can now use SAP GUI to establish a new local connection towards your SAP BW system.

To set up a local SAP HANA connection for Design Studio, follow these steps:

1. Start the Design Studio designer.

2. Navigate to the menu Tools > Preferences (see Figure 4.2).

3. Select the area Application Design.

4. Select the entry Backend Connections (see Figure 4.3).

5.   In the SAP HANA Connections (HTTP) area click Add (see Figure 4.4).

Figure 4.4 SAP HANA connection details

6.   You can enter the details according to Table 4.4.

Field Name	Value Description
Name	The name of the connection the user sees in the BI tool.
Host	The fully qualified hostname of your SAP HANA system.
Port	The default port would be 80<Instance Number>, so for an instance number 00 the port would be 8000.
Protocol	Either HTTP or HTTPS.

Table 4.4 Connection details

7.   Enter the required details and click OK.

8.   The new connection is added to the list of local connections and you will now be able to use the local connection with Design Studio.

In the next section we will create our first dashboard using Design Studio.

CHAPTER 4

4.5 Creating Your First Dashboard in Design Studio

In the following steps you will learn how to use some of the basic functionality of Design Studio. You will leverage the connection created previously in the Central Management Console (CMC) of your SAP BusinessObjects BI platform system. For our example steps we are using a set of BEx Queries based on the SAP NetWeaver Demo model (http://scn.sap.com/docs/DOC-8941) with the details shown in the images below.

Our first dashboard will show a set of key performance indicators (KPIs) after the user selects a specific customer from a drop-down list and a chart that shows the trend over the last 12 months. Figure 4.5 shows the first query in the BEx Query designer, returning the key figures broken down by customer (Sold-to Party).

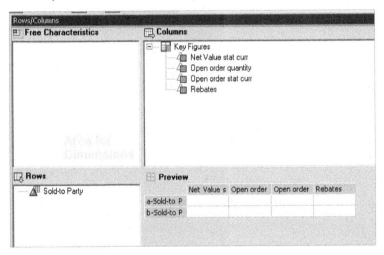

Figure 4.5 BEx Query

In addition to these elements, the BEx Query also does contain an optional variable for the characteristic Customer. We will use the first BEx Query to retrieve the KPIs that are shown per customer. The BEx Query is based on the MultiProvider 0D_NW_M01 of the SAP NetWeaver Demo model. Figure 4.6 shows the second query, which will return the key figures broken down by the Calendar month.

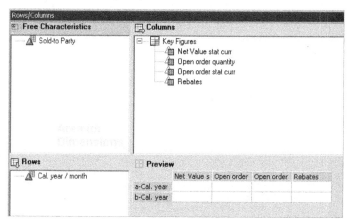

Figure 4.6 BEx Query

In addition, the query contains the customer (Sold-to Party) in the Free Characteristics area, so that we can filter the information per customer. The second query also contains an SAP EXIT variable that will filter the data to the 12 months of the current year.

The third BEx Query is shown in Figure 4.7.

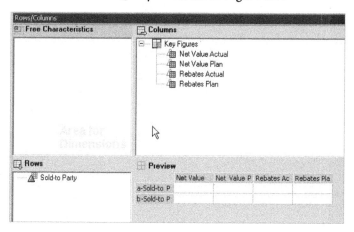

Figure 4.7 BEx Query

It will deliver the actual and plan values for the Net Value and for the Rebates amount broken down by customer (Sold-to Party). The key figures in the Columns are restricted key figures, which are using the characteristic Value Type to make a distinction between the actual and plan values. We will use this query to show the KPI tracking and the forecast revenue value in the top area of the dashboard.

We will use these three BEx Queries to create the dashboard now.

1. Start Design Studio.

2. Navigate to the menu Application > New (see Figure 4.8).

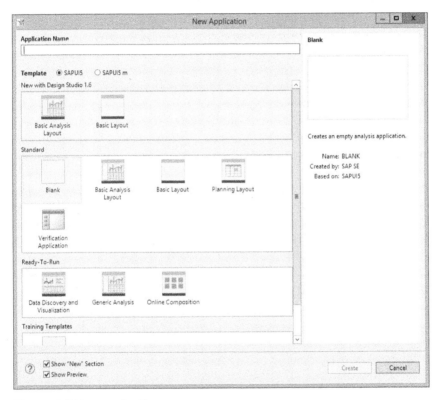

Figure 4.8 New application

CHAPTER 4

3.   Enter an Application Name.

4.   Select the option SAPUI5 m.

5.   Select the Standard template Blank.

6.   Click Create.

7.   Navigate to the Outline panel and select the folder Data Source.

8.   Right-click and select the menu Add Data Source... (see Figure 4.9).

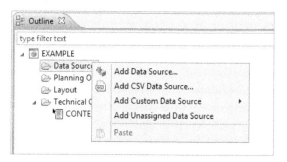

Figure 4.9 Add a Data Source

9.   You are presented with a new screen (see Figure 4.10) to establish the connection.

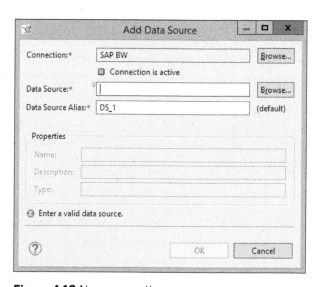

Figure 4.10 New connection

10. Click Browse for the option Connection.

11. Select the connection to the previously defined BEx Queries.

12. Click OK.

13. Click Browse for the option Data Source.

14. Click the tab Search.

15. Search for the first query based on the technical name.

16. Select the query and click OK.

17. Enter KEYFIGURES_BY_CUSTOMER as the Data Source Alias.

18. Click OK.

19. Navigate to the Outline panel and select the folder Data Source.

20. Right-click and select the menu Add Data Source.

21. Click Browse for the option Connection.

22. Select the connection to the previously defined BEx Queries.

23. Click OK.

24. Click Browse for the option Data Source.

25. Click the tab Search.

26. Search for the second query based on the technical name.

27. Select the query and click OK.

28. Enter TREND_PER_CUSTOMER as the Data Source Alias.

29. Click OK.

30. Right-click the data source TREND_PER_CUSTOMER.

31. Select the menu Edit Initial View.

32. With a simple drag-and-drop navigation, move the characteristic Calendar Year/Month to the Columns and Measures to the Rows.

33. Click OK.

34. Navigate to the Outline panel and select the folder Data Source.

35. Right-click and select the menu Add Data Source.

36. Click Browse for the option Connection.

37. Select the connection to the previously defined BEx Queries.

CHAPTER 4

38. Click OK.

39. Click Browse for the option Data Source.

40. Click the tab Search.

41. Search for the third query based on the technical name.

42. Select the query and click OK.

43. Enter ACTUAL_PLAN_PER_CUSTOMER as the Data Source Alias.

44. Click OK.

45. Now add a Dropdown Box from the Basic Components to your empty canvas of the new application.

46. Navigate to the Properties shown on the right side (see Figure 4.11).

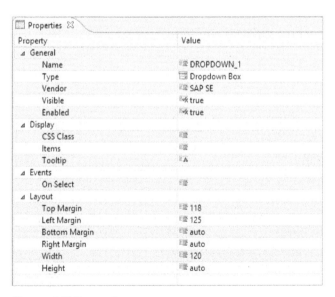

Property	Value
⊿ General	
Name	DROPDOWN_1
Type	Dropdown Box
Vendor	SAP SE
Visible	true
Enabled	true
⊿ Display	
CSS Class	
Items	
Tooltip	
⊿ Events	
On Select	
⊿ Layout	
Top Margin	118
Left Margin	125
Bottom Margin	auto
Right Margin	auto
Width	120
Height	auto

Figure 4.11 Properties

47. Configure the values as shown below:
    - Top Margin: 10
    - Left Margin: 700
    - Bottom Margin: auto
    - Right Margin: auto
    - Width: 300
    - Height: auto

48. Navigate to the property Name and enter DROPDOWN_ CUSTOMER as the name for the Dropdown Box.

So far our application has the data sources and the Dropdown Box. In the next steps we will add text boxes, which we then will use to show the values per customer.

49. Now add four text boxes from the Basic Components to your new application.

50. Select the first text box.

51. Open the Properties for the first text box and configure the following values:
    - Top Margin: 100
    - Left Margin: 50
    - Bottom Margin: auto
    - Right Margin: auto
    - Width: 150
    - Height: 30

52. Navigate to the property Name and enter NET_VALUE_TITLE.

53. Navigate to the property Text and enter Total Revenue.

54. Navigate to the property CSS Style for the text box.

55. Click the button for the CSS Style.

CHAPTER 4

56. Enter the following values:

    background-color: #D9D9D9;
    color: #000000;
    padding: 5px;
    font-family: arial;
    font-style: normal;
    font-weight: bold;
    font-size: 20px;
    text-align: center;

57. Open the Properties for the second text box and configure the following values:

    - Top Margin: 100
    - Left Margin: 220
    - Bottom Margin: auto
    - Right Margin: auto
    - Width: 150
    - Height: 30

58. Navigate to the property Name and enter OPEN_ORDERS_QTY_TITLE.

59. Navigate to the property Text and enter Open Orders.

60. Navigate to the property CSS Style for the text box.

61. Click the button for the CSS Style.

62. Enter the following values:

    background-color: #D9D9D9;
    color: #000000;
    padding: 5px;
    font-family: arial;
    font-style: normal;
    font-weight: bold;
    font-size: 20px;
    text-align: center;

63. Open the Properties for the third text box and configure the following values:

- Top Margin: 100
- Left Margin: 390
- Bottom Margin: auto
- Right Margin: auto
- Width: 150
- Height: 30

64. Navigate to the property Name and enter REBATES_TITLE.

65. Navigate to the property TEXT and enter Rebates.

66. Navigate to the property CSS Style for the text box.

67. Click the button for the CSS Style.

68. Enter the following values:

    background-color: #D9D9D9;
    color: #000000;
    padding: 5px;
    font-family: arial;
    font-style: normal;
    font-weight: bold;
    font-size: 20px;
    text-align: center;

69. Open the Properties for the fourth text box and configure the following values:

- Top Margin: 100
- Left Margin: 560
- Bottom Margin: auto
- Right Margin: auto
- Width: 150
- Height: 30

CHAPTER 4

70. Navigate to the property Name and enter FORECAST_TITLE.

71. Navigate to the property Text and enter Forecast.

72. Navigate to the property CSS Style for the text box.

73. Click the button for the CSS Style.

74. Enter the following values:

    background-color: #D9D9D9;
    color: #000000;
    padding: 5px;
    font-family: arial;
    font-style: normal;
    font-weight: bold;
    font-size: 20px;
    text-align: center;

At this point we have added four text boxes with the headings for our KPIs to the canvas (see Figure 4.12). In addition to these headings we are now going to add another set of four text boxes, which we will use to show the values per customer.

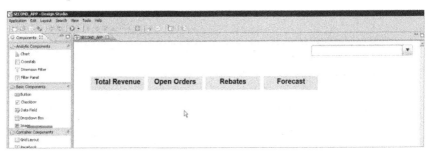

Figure 4.12 Design Studio application

75. Now add four more text boxes from the Basic Components to your new application.

76. Select the first of the newly added text boxes.

77. Open the Properties for the first new text box and configure the following values:

- Top Margin: 140

- Left Margin: 50

- Bottom Margin: auto

- Right Margin: auto

- Width: 150

- Height: 30

78. Navigate to the property Name and enter NET_VALUE_VALUE.

79. Navigate to the property CSS Style for the text box.

80. Click the button for the CSS Style.

81. Enter the following values:

    background-color: #D9D9D9;
    color: #000000;
    padding: 5px;
    font-family: arial;
    font-style: normal;
    font-weight: bold;
    font-size: 20px;
    text-align: center;

82. Navigate to the property Text and remove the default text.

83. Open the Properties for the second new text box and configure the following values:

- Top Margin: 140

- Left Margin: 220

- Bottom Margin: auto

- Right Margin: auto

- Width: 150

- Height: 30

CHAPTER 4

84. Navigate to the property Name and enter OPEN_ORDERS_QTY_VALUE.

85. Navigate to the property CSS Style for the text box.

86. Click the button for the CSS Style.

87. Enter the following values:

background-color: #D9D9D9;
color: #000000;
padding: 5px;
font-family: arial;
font-style: normal;
font-weight: bold;
font-size: 20px;
text-align: center;

88. Navigate to the property Text and remove the default text.

89. Open the Properties for the third new text box and configure the following values:

- Top Margin: 140
- Left Margin: 390
- Bottom Margin: auto
- Right Margin: auto
- Width: 150
- Height: 30

90. Navigate to the property Name and enter REBATES_VALUE.

91. Navigate to the property CSS Style for the text box.

92. Click the button for the CSS Style.

93. Enter the following values:

background-color: #D9D9D9;
color: #000000;
padding: 5px;
font-family: arial;
font-style: normal;
font-weight: bold;

font-size: 20px;
text-align: center;

94. Navigate to the property Text and remove the default text.

95. Open the Properties for the fourth new text box and configure the following values:

    - Top Margin: 140
    - Left Margin: 560
    - Bottom Margin: auto
    - Right Margin: auto
    - Width: 150
    - Height: 30

96. Navigate to the property Name and enter FORECAST_VALUE.

97. Navigate to the property CSS Style for the text box.

98. Click the button for the CSS Style.

99. Enter the following values:

    background-color: #D9D9D9;
    color: #000000;
    padding: 5px;
    font-family: arial;
    font-style: normal;
    font-weight: bold;
    font-size: 20px;
    text-align: center;

100. Navigate to the property Text and remove the default text.

101. Add a chart from the Analytical Components to the lower-left part of your application.

102. Open the Properties for the chart and configure the following values:

    - Top Margin: 250
    - Left Margin: 50
    - Bottom Margin: auto

- Right Margin: auto
- Width: 700
- Height: 400

103. Now drag and drop the data source TREND_PER_CUSTOMER to the chart. The chart should show the actual data now.

104. In the Properties of the chart, set the Chart Type to the value Line.

105. Select the top item in the Outline panel representing your application (see Figure 4.13).

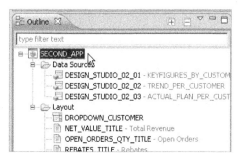

Figure 4.13 Application object

106. Navigate to the Properties for the application.

107. In the Properties click the button for the On Startup property.

108. Click the button to open the script editor.

109. Add the following script:

```
DROPDOWN_CUSTOMER.setItems(KEYFIGURES_
BY_CUSTOMER.getMemberList("0D_NW_SOLD",
MemberPresentation.INTERNAL_KEY, MemberDisplay.TEXT,
100));
```

110. The script will load the members from characteristic Sold-To Party (0D_NW_SOLD) into the drop-down list.

111. Select the Dropdown Box in your application.

112. In the Properties click the button for the On Select property.

113. Add the following script (Figure 4.14):

```
APPLICATION.setVariableValue("VAR_CUSTOMER", DROPDOWN_CUSTOMER.
getSelectedValue());
TREND_PER_CUSTOMER.setFilter("0D_NW_SOLD", DROPDOWN_CUSTOMER.
getSelectedValue());
ACTUAL_PLAN_PER_CUSTOMER.setFilter("0D_NW_SOLD", DROPDOWN_
CUSTOMER.getSelectedValue());
NET_VALUE_VALUE.setText(KEYFIGURES_BY_CUSTOMER.
getDataAsString("12RTRMIZYBHKR57262FQSIAF8", {
        "0D_NW_SOLD": DROPDOWN_CUSTOMER.getSelectedValue()
}));
OPEN_ORDERS_QTY_VALUE.setText(KEYFIGURES_BY_CUSTOMER.
getDataAsString("12RTRMIZYBHKR57262FQSIGQS", {
        "0D_NW_SOLD": DROPDOWN_CUSTOMER.getSelectedValue()
}));
REBATES_VALUE.setText(KEYFIGURES_BY_CUSTOMER.
getDataAsString("12RTRMIZYBHKR57262FQSITDW", {
        "0D_NW_SOLD": DROPDOWN_CUSTOMER.getSelectedValue()
}));
FORECAST_VALUE.setText(KEYFIGURES_BY_CUSTOMER.
getDataAsString("12RTRMIZYBHKR57262FQSIAF8", {
        "0D_NW_SOLD": DROPDOWN_CUSTOMER.getSelectedValue()
}));
```

Figure 4.14 Add script

Before we continue, let's clarify the elements of the script.

```
APPLICATION.setVariableValue("VAR_CUSTOMER", DROPDOWN_
CUSTOMER.getSelectedValue());
```

The first part of the script is using the selected value from the drop down and passes the value to the BEx Query variable in the first query to filter the data set to the selected customer.

```
TREND_PER_CUSTOMER.setFilter("0D_NW_SOLD", DROPDOWN_
CUSTOMER.getSelectedValue());

ACTUAL_PLAN_PER_CUSTOMER.setFilter("0D_NW_SOLD",
DROPDOWN_CUSTOMER.getSelectedValue());
```

CHAPTER 4

The next two lines in the script are also filtering the data, in this case the second and third data sources, based on the selected value from the drop down. In this case the filtering is not done using a BEx Query variable, but instead the filter value is just passed to the characteristic.

```
NET_VALUE_VALUE.setText(KEYFIGURES_BY_CUSTOMER.
getDataAsString("12RTRMIZYBHKR57262FQSIAF8", { "0D_NW_
SOLD": DROPDOWN_CUSTOMER.getSelectedValue() }));
```

The rest of the script is using the getDataString and setText methods, to first get a single value from the data source by specifying which key figure and which filter values (for example, "0D_NW_SOLD": DROPDOWN_CUSTOMER.getSelectedValue()), and then to use the received value and set it as text for the text boxes we added to the canvas.

Basically, the script will filter the data sources we added. It will receive the four needed values from the BEx Queries and display those values in the text boxes (see Figure 4.15).

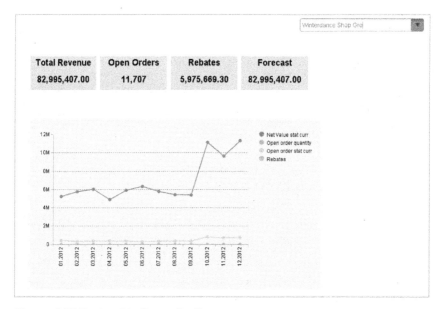

Figure 4.15 Design Studio application

At this point our dashboard does display the set of KPIs and the chart per customer. The data will change each time the user selects a different customer from the drop down.

Reference of Key Figures

In the script above, you can identify that the used key figures are referenced by their technical ID (a GUID) instead of the technical name. Characteristics are referred to by their technical names and key figures are referred to by using the GUID. The easiest way to retrieve the correct technical names in the scripting is to leverage the CTRL+Space key combination to receive a list of valid entries and to simply select the needed key figure from the list.

114. Select the menu Application > Save All.

115. Select the menu Application > Execute Locally.

In this section you learned some basic dashboard building steps in Design Studio. In the next chapter we will review the details of SAP BusinessObjects Lumira in combination with your SAP landscape.

4.6 Summary

In this chapter you learned how you can leverage Design Studio in combination with data from your SAP BW and SAP HANA systems. You also learned how the product supports the existing metadata from your SAP BW and your SAP HANA systems. In the next chapter you will learn more about SAP Lumira.

CHAPTER 4

CHAPTER 5

SAP BusinessObjects Lumira

In this chapter you will learn about the data-connectivity options and the level of support for the metadata in SAP BW and SAP HANA, and you will create your first report using SAP BusinessObjects Lumira.

5.1 Data-Connectivity Overview

Figure 5.1 shows the data connectivity between SAP Lumira and your SAP BW, SAP ERP, and SAP HANA systems.

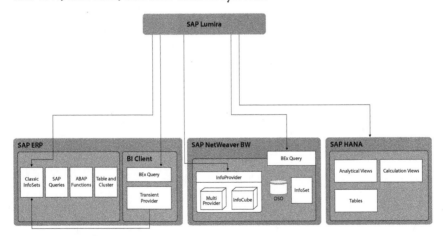

Figure 5.1 Data connectivity

SAP Lumira offers direct connectivity towards SAP HANA, SAP BW, and SAP ERP. In regards to data connectivity, SAP Lumira is able to:

- Establish a direct connectivity to classic InfoSets in your SAP ERP system.

- Connect to your BEx Queries as well as InfoProviders in your SAP BW system. SAP Lumira offers only a download from SAP BW at this point in time (August 2016) and not an online connectivity option.

- Connect directly to your SAP HANA system. SAP Lumira offers the option to connect online to SAP HANA or to download the information to SAP Lumira.

In addition to direct connectivity, SAP Lumira is also able to leverage the Universe layer (see Figure 5.2).

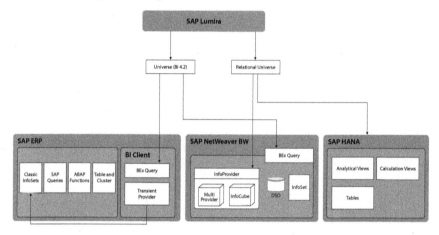

Figure 5.2 SAP Lumira and Universes

Using the Universe layer, SAP Lumira is able to offer additional connectivity options:

- A relational Universe can be used to set up connectivity to SAP HANA models and tables.

- A relational Universe can be established towards InfoProviders in SAP BW.
- Starting with SAP BusinessObjects BI 4.2, a Universe can also be created on top of BEx Queries.

In addition to the connectivity options and the ability to use the information from your SAP ERP, SAP BW, and SAP HANA systems, SAP Lumira also offers the ability to download the data from SAP HANA and SAP BW to your desktop so that you can still use the data without the need for an online connection to your server.

In this section we reviewed the different data-connectivity options available with SAP Lumira. In the next section you will learn about the level of support of your existing metadata from SAP BW in combination with SAP Lumira.

5.2 Supported and Unsupported SAP BW Elements

In this section we review the level of support for your existing metadata inside the SAP BW system for SAP Lumira. Table 5.1 shows which objects are supported when using SAP Lumira as the BI client tool.

SAP BW Metadata	Direct Access
Direct access to InfoCubes and MultiProviders	Yes
Access to BEx Queries	Yes
Characteristic Values	
Key	Yes
Short description	Yes
Medium and long descriptions	Yes
BEx Query Features	
Support for hierarchies	Limited
Support for free characteristics	Yes
Support for calculated and restricted key figures	Yes
Support for currencies and units	No

Table 5.1 Supported and unsupported BEx Query features for SAP Lumira *(continues)*

CHAPTER 5

193

SAP BW Metadata	Direct Access
Support for custom structures	Yes
Support for formulas and selections	Yes
Support for filters	Yes
Support for display and navigational attributes	Yes
Support for conditions in rows	Yes
Support for conditions in columns	No
Support for conditions for fixed characteristics	Yes
Support for exceptions	No
Compounded characteristics	Yes
Constant selection	No
Default values in BEx Query	No
Number scaling factor	No
Number of decimals	No
Calculate rows as (local calculation)	No
Sorting	No
Hide/Unhide	No
Display as hierarchy	No
Reverse sign	Yes
Support for reading master data	Yes
Data Types	
Support for CHAR (characteristics)	Yes
Support for NUMC (characteristics)	No
Support for DATS (characteristics)	Yes
Support for TIMS (characteristics)	Yes
Support for numeric key figures such as Amount and Quantity	Yes
Support for Date (key figures)	Yes
Support for Time (key figures)	Yes
BEx Variables — Processing Type	
User input	Yes
Authorization	Yes
Replacement path	Yes

Table 5.1 Supported and unsupported BEx Query features for SAP Lumira *(continues)*

CHAPTER 5

SAP BW Metadata	Direct Access
SAP exit/custom exit	Yes
Precalculated value set	Yes
General Features for Variables	
Support for optional and mandatory variables	Yes
Support for key date dependencies	Yes
Support for default values	Yes
Support for variable variants	No
Support for personalized values	No
BEx Variables—Variable Type	
Single value	Yes
Multi-single value	Yes
Interval value	Yes
Selection option	Yes
Hierarchy variable	No
Hierarchy node variable	Yes
Hierarchy version variable	Yes
Text variable	Yes
EXIT variable	Yes
Single key date variable	Yes
Multiple key dates	Yes
Formula variable	Yes

Table 5.1 Supported and unsupported BEx Query features for SAP Lumira *(continued)*

CHAPTER 5

In Table 5.2 you can see how the direct-access method uses elements from the BEx Query and how the objects are mapped to SAP Lumira.

BEx Query Element	SAP Lumira
Characteristic	For each characteristic you'll receive a field, and when configuring the data acquisition, you need to decide which elements—Key and Text—for each characteristic should be retrieved.
Hierarchy	Each available hierarchy is shown as an external hierarchy and can be retrieved as part of the data acquisition. During the configuration of the data acquisition you need to define the number of levels for each hierarchy that needs to be retrieved.
Key figure	Each selected key figure is retrieved, but you are missing additional information on unit or currency information.
Calculated/Restricted key figure	Each calculated and restricted key figure is treated like a key figure. The user does not have access to the underlying definition in SAP Lumira.
Filter	Filters are applied to the underlying query and you do not have access to those filter values in SAP Lumira.
Display attribute	Display attributes become standard fields and can be included into result sets in combination with the dimension itself.
Navigational attribute	Navigational attributes are treated the same way as characteristics.
Variable	Each variable with the property Ready for Input results in a prompt. During data acquisition you still have the option to define the input values.
Custom structure	A custom structure is available as an element during the data acquisition and is treated like a characteristic.

Table 5.2 BEx Query metadata mapping for SAP Lumira

As you can see in Table 5.1 and Table 5.2, SAP Lumira is lacking some important elements when it comes to SAP BW metadata, such as the Unit and Currency information, as well as several display-specific settings such as a scaling factor. In addition, the support for hierarchies is limited and the user creating a new connection towards SAP BW has to decide—at the point of acquiring the data—how many levels of the hierarchy will be retrieved. This cannot be changed dynamically.

You will also notice that variables from the underlying BEx Query are available for use during data acquisition, but when refreshing an SAP Lumira document the user will not be prompted for new input values for those BEx variables.

After reviewing the support for SAP BW elements, we will review the support for SAP HANA elements in the next section.

5.3 Supported and Unsupported SAP HANA Elements

In this section we review the level of support for your existing metadata inside SAP HANA for SAP Lumira. Table 5.3 shows which of the objects are supported when using SAP Lumira as the BI client tool.

SAP HANA Metadata	Direct Access to SAP HANA
Analytical Model	Yes
Calculation Model	Yes
Attribute Model	Yes
Dimension Key	Yes
Dimension description (Label Column)	Yes
Unit Currency	No
Calculated column	Yes
Restricted column	Yes
Variable	Yes
Input Parameter	Yes
Level-Based Hierarchy	Yes
Parent-Child Hierarchy	No

Table 5.3 Supported elements from SAP HANA

After reviewing the support for your SAP HANA-based metadata, we will learn how to set up the local connections in SAP Lumira.

CHAPTER 5

5.4 Setting Up Local Connections

SAP Lumira is able to leverage connections from the SAP BusinessObjects BI platform as well as establish location connections towards SAP HANA or SAP BW. In this section we will first set up a local connection towards SAP BW and then set up a local connection towards SAP HANA. In general, SAP Lumira does not provide the ability to create a list of connections locally, so the local connection is established each time you create content using SAP Lumira. In case of a local connection for SAP BW, SAP Lumira leverages the entries from SAP GUI.

To establish a local connection to SAP BW, follow these steps:

1. Start SAP Lumira.

2. Open the menu File > Extensions.

3. Ensure that the SAP BW Data Acquisition Connector is installed.

4. Close the screen for the Extensions.

5. Select the menu File > New (see Figure 5.3).

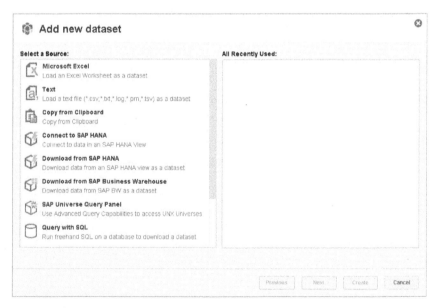

Figure 5.3 New SAP Lumira content

6. Select the entry Download from SAP Business Warehouse.

7. Click Next (see Figure 5.4).

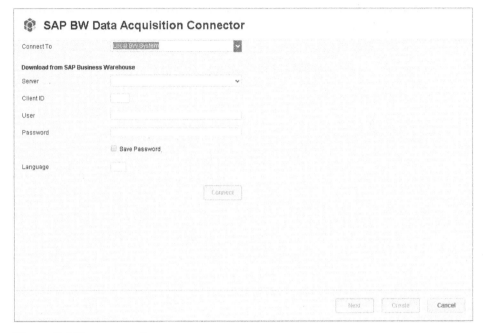

Figure 5.4 New connection

8. You can now choose between a local connection or a connection from the SAP BusinessObjects BI platform by setting the option for the Connect To field.

9. Set the option to Local BW System.

CHAPTER 5

10. Open the list of entries for the Server option (see Figure 5.5).

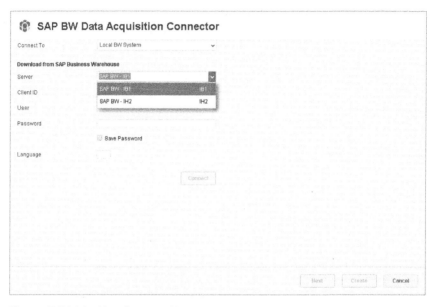

Figure 5.5 List of local connections

11. SAP Lumira uses the entries from SAP GUI and, as you can see, there is an option to define a new local connection directly in SAP Lumira, but instead SAP Lumira offers the defined server entries from SAP GUI.

You can now choose from the list of entries and establish the connection to your SAP BW system. We will use the connection in the next section for creating our first content with SAP Lumira.

To establish a local connection to SAP HANA, follow these steps:

1. Start SAP Lumira.
2. Select the menu File > New (see Figure 5.3).
3. Select the entry Connect to SAP HANA.
4. Click Next (see Figure 5.6).

Figure 5.6 SAP HANA connection

5. You can now choose between a local connection or a connection from the SAP BusinessObjects BI platform.

6. Set the Connect To option to HANA.

7. You can then enter the Server, Port, and Instance. By default, the Port will be 3<Instance number>15, so for an SAP HANA Instance number 00, the default Port would be 30015.

After a successful connection, the connection details will be available in the list of servers the next time.

In this section we learned how to establish a local connection towards SAP HANA and SAP BW using SAP Lumira. In the next section we will create our first content using SAP Lumira combined with SAP BW.

CHAPTER 5

5.5 Creating Your First Data Discovery Content in SAP Lumira

In the following steps you will learn how to use some of the basic functionality of SAP Lumira. You will leverage the connection created previously in the Central Management Console (CMC) of your SAP BusinessObjects BI platform system.

For our example steps we are using a BEx Query based on the SAP NetWeaver Demo model (http://scn.sap.com/docs/DOC-8941) with the following details:

Rows:

- Product Group (0D_NW_PROD__0D_NW_PRDGP)

Columns:

- Net Value (0D_NW_NETV)
- Costs (0D_NW_COSTV)

Free Characteristics:

- Country (0D_NW_CNTRY)
- Region (0D_NW_REGIO)
- Product Category (0D_NW_PROD__0D_NW_PRDCT)
- Cal. Year/Month (0CALMONTH)
- Calendar Year (0CALYEAR)

To create your first SAP Lumira document, follow these steps:

1. Start SAP Lumira.

2. In SAP Lumira select the menu File > New (see Figure 5.7).

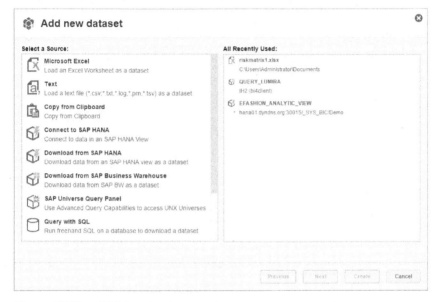

Figure 5.7 New SAP Lumira document

3. Select the option Download from SAP Business Warehouse.

4. Click Next.

5. Set the Connect To option to the value SAP BusinessObjects BI Platform (see Figure 5.8).

Figure 5.8 Data connection

6. Enter your system details and user credentials.

7. Click Connect.

8. Select the previously created connection to your SAP BW system.

9. Click Next (see Figure 5.9).

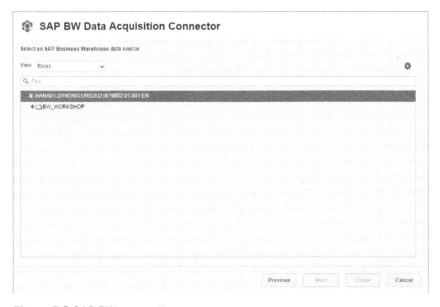

Figure 5.9 SAP BW connection

10. Use the Find option to search for your BEx Query.

11. Select the BEx Query.

12. Click Next (see Figure 5.10).

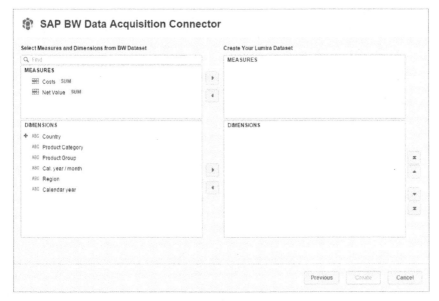

Figure 5.10 BEx Query details

CHAPTER 5

13. Using the arrows in the middle, move all the measures into the Measures area on the right side.

14. Using the arrows in the middle, move all the entries in the Dimensions list to the Dimensions area on the right side (see Figure 5.11).

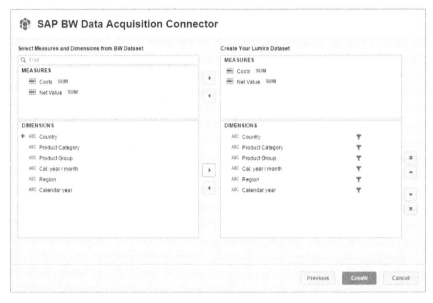

Figure 5.11 Data selection

15. Now select dimension Cal. Year/Month in the right area.

16. Click the Settings icon (circle icon above the letter P) on the right side (see Figure 5.12).

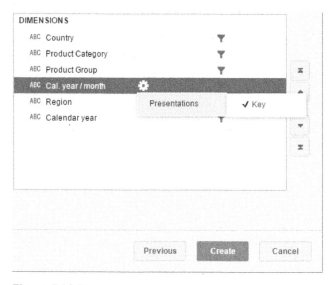

Figure 5.12 Dimension settings

17. Select the menu **Presentations > Key**.

18. Click **Create** (see Figure 5.13).

Figure 5.13 SAP Lumira

19. Click Prepare (see Figure 5.14).

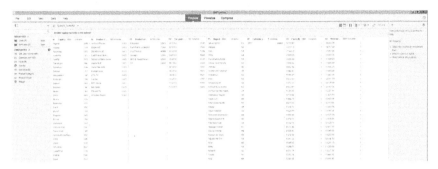

Figure 5.14 The Prepare Room

You can now start making some adjustments to the actual data set.

20. Select the dimension Country in the list of Dimensions on the left side.

21. Use the Settings icon to see the list of options (see Figure 5.15).

Figure 5.15 Settings menu

22. Select the menu entry Create a geographic hierarchy and then click the by Names option (see Figure 5.16).

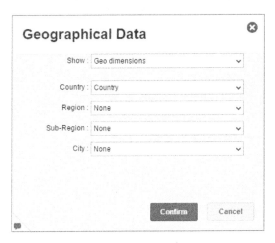

Figure 5.16 Geographic hierarchy

23. Set the option Country to dimension Country.

24. Set the option Region to dimension Region.

25. Click Confirm (see Figure 5.17).

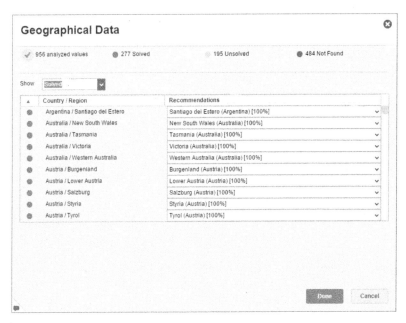

Figure 5.17 Geographical data

26. You are presented with the list of solved and unsolved Countries and Regions. (Solved means the system found the items.) For our example we will continue with the default matches.

27. Click Done.

28. Now select the column header for the dimension Cal. Year/Month.

29. We would now like to separate the actual month from the year. In the Data actions area on the right side, select the option Split (see Figure 5.18).

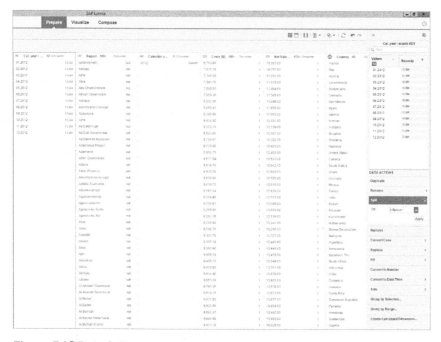

Figure 5.18 Data Actions

30. For the option On replace the value <Space> with a "." (dot).

31. Click Apply.

32. The system now generates two additional columns with the separated values for the Calendar Year and Calendar Month.

33. Navigate back to the column Cal. Year/Month and select the column header.

34. In the top-right corner of the column header click the Settings icon (see Figure 5.19).

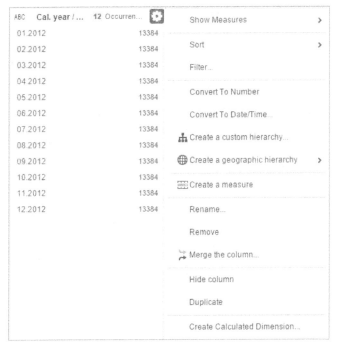

Figure 5.19 Settings menu

35. Select the menu Hide column.

36. Now navigate to the two new columns we just created by splitting the value. These columns should be named Cal. Year/Month Key 2 and Cal. Year/Month Key 3.

37. Select the column header of the column representing the Month Value (Cal. Year/Month Key 2).

38. Open the Settings menu in the top-right corner and select the menu option Rename.

39. Enter Calendar Month as the new Name.

40. Select the column header of the column representing the Calendar Year (Cal. Year/Month Key 3).

41. Open the Settings menu in the top-right corner and select the menu option Hide column.

42. Select the measure Net Value [$] and open the menu Create Calculated Measure... (see Figure 5.20).

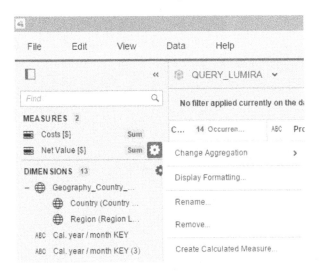

Figure 5.20 Calculated Measure

43. Enter Profit as the Measure Name.

44. Enter the following code in the Formula box (see Figure 5.21):

    {Net Value [$]} - {Costs [$]}

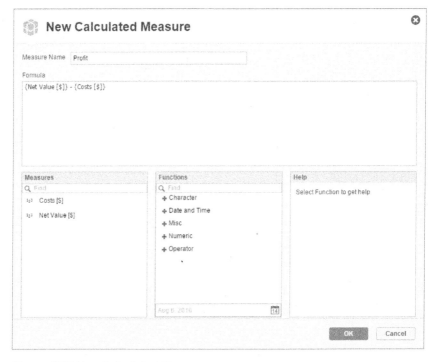

Figure 5.21 New Calculated Measure

45. Click OK.

46. Select the measure Profit and open the Settings menu and select the menu entry Create Calculated Measure...

47. This time enter Profit in % as the Measure Name.

48. As the Formula, enter the following:

    {Profit} / {Net Value [$]}

49. Click OK.

50. Open the Settings menu for measure Profit in % (see Figure 5.22).

Figure 5.22 Settings menu

51. Select the menu Display Formatting... (see Figure 5.23).

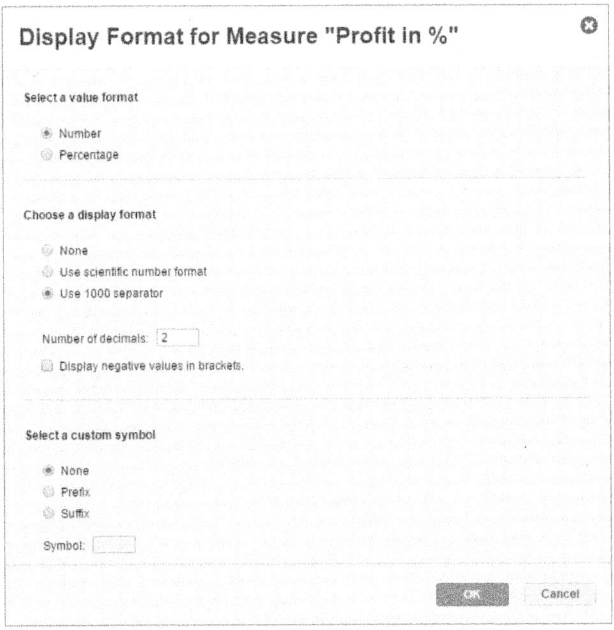

Display Format for Measure "Profit in %"

Select a value format

- Number
- Percentage

Choose a display format

- None
- Use scientific number format
- Use 1000 separator

Number of decimals: 2

- Display negative values in brackets.

Select a custom symbol

- None
- Prefix
- Suffix

Symbol:

OK   Cancel

Figure 5.23 Display format

52. Set the Value Format to Percentage.

53. Set the Number of decimals to 0.

54. Click OK.

55. Navigate to the Visualize Room (see Figure 5.24).

Figure 5.24 Visualize Room

56. In the Visualize area, choose a Bar Chart (see Figure 5.25).

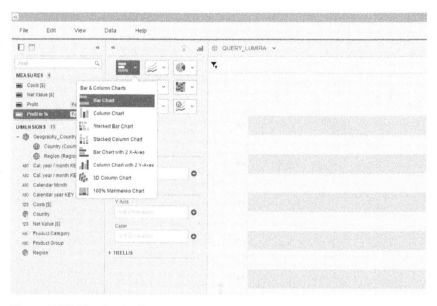

Figure 5.25 Chart selection

57. Using a drag-and-drop navigation, add the dimension Product Group to the Y-Axis.

58. Using a drag-and-drop navigation, add the measure Net Value to the X-Axis (see Figure 5.26).

Figure 5.26 Bar chart

59. As we are only interested in the Top customers we will rank the information.

60. In the toolbar above the chart, select the option for Ranking (see Figure 5.27).

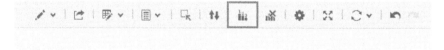

Figure 5.27 Chart menu

61. Set the ranking to be based on the measure Net Value.

62. Select the option Top 5.

63. Set the last option to Product Group (see Figure 5.28).

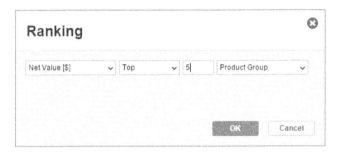

Figure 5.28 Ranking

64. Click OK (see Figure 5.29).

Figure 5.29 Ranked chart

65. In the gallery, click the thumbnail of the chart and select the menu Duplicate (see Figure 5.30).

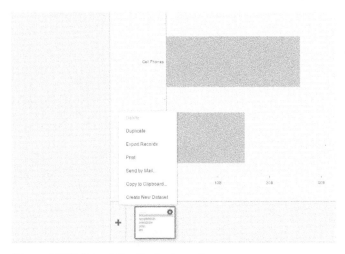

Figure 5.30 Select Duplicate from the menu

66. The option Duplicate allows you to create a copy of your chart so that you can go back to the current chart definition at a later point in time.

67. We are now going to take a look at the data from a product perspective.

CHAPTER 5

68. Set the Chart Type to Geo Choropleth Chart (see Figure 5.31).

Figure 5.31 Select the visualization type

69. Using a drag-and-drop navigation, place the geographic hierarchy for dimension Country that we created previously to the option Geography (see Figure 5.32).

Figure 5.32 Geographic hierarchy

70. Add the measure Net Value to the Color option (see Figure 5.33).

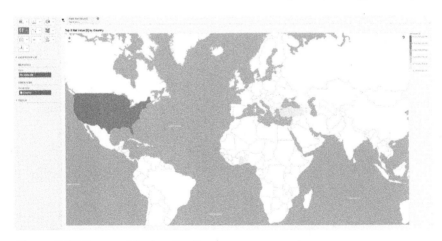

Figure 5.33 Geographic visualization

71. Now navigate to the dimension Country and use the Settings menu to change the geographic hierarchy to the Region level (see Figure 5.34).

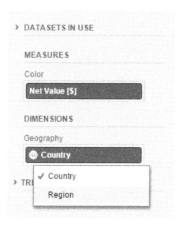

Figure 5.34 Geographic level

72. Now the map displays the next level of our geographic hierarchy (see Figure 5.35).

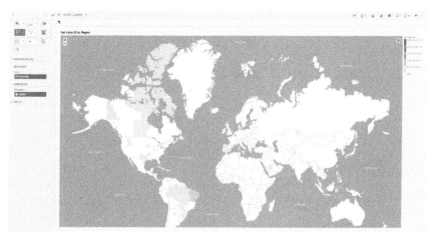

Figure 5.35 Regional level

73. Now change the chart type to Geo Pie Chart.

74. With a drag-and-drop navigation, move the dimension Product Group to the field Color (see Figure 5.36).

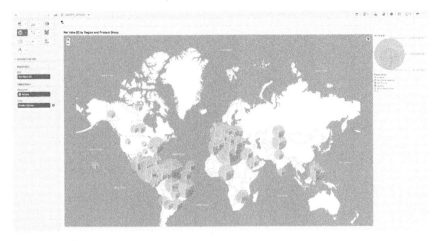

Figure 5.36 Geo Pie Chart view

75. Using the +/– icon on the top-left corner of the map, zoom into the details of the map (see Figure 5.37).

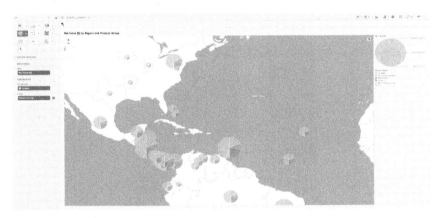

Figure 5.37 Map zoom

76. Select the menu File > Save as to save your SAP Lumira document (see Figure 5.38).

CHAPTER 5

Figure 5.38 Save options

77. You have the option to save your document locally or to your SAP BusinessObjects BI platform.

78. Select the entry SAP BI Platform (see Figure 5.39).

Figure 5.39 Save to the SAP BusinessObjects BI platform

79. For the URL for the field Log in to SAP BusinessObjects BI Platform you have to enter the name of your Java Application Server and the port number. This URL is the RESTful Web Service URL of your SAP BusinessObjects BI system and the default port is 6405.

80. Enter your user credentials and the corresponding Authentication Type.

81. Click Connect (see Figure 5.40).

Figure 5.40 Save to the BI platform

82. Select a folder and enter a name for your SAP Lumira document.

83. Click Save.

84. Now launch the SAP BusinessObjects BI platform Java BI launch pad.

85. Log on with your credentials.

86. Navigate to the folder where you stored your SAP Lumira document.

87. Select your SAP Lumira document.

88. Either double-click or open the context menu (right-click) and select View (see Figure 5.41).

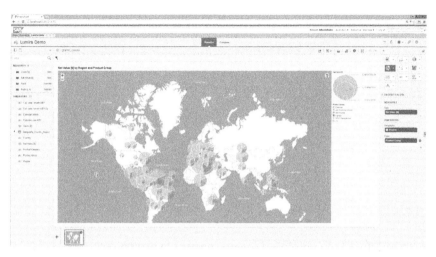

Figure 5.41 SAP Lumira document

You can now share your SAP Lumira document with your colleagues using the SAP BusinessObjects BI platform.

5.6 Summary

In this chapter you learned how to leverage SAP Lumira in combination with data from your SAP BW and SAP HANA systems. You also learned how the product supports the existing metadata from your SAP BW system and your SAP HANA system. In the next chapter you will learn more about Web Intelligence and Universes.

SAP BusinessObjects Web Intelligence (Web Intelligence)

In this chapter you will learn about the data-connectivity options and the level of support for the metadata in SAP BW and SAP HANA using Web Intelligence. We will review direct connectivity as well as the ability to leverage the Semantic Layer of the SAP BusinessObjects BI platform.

6.1 Data-Connectivity Overview

Figure 6.1 shows the data connectivity between Web Intelligence and your SAP BW, SAP ERP, and SAP HANA systems.

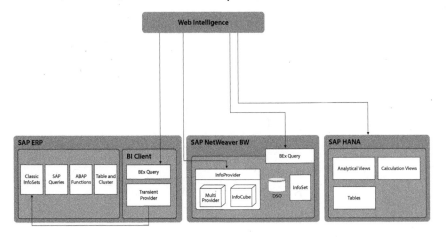

Figure 6.1 Data connectivity

Web Intelligence offers direct connectivity towards SAP HANA and SAP BW. In regards to data connectivity, Web Intelligence is able to:

- Connect to your BEx Queries as well as InfoProviders in your SAP BW system.

- Connect directly to your SAP HANA system. Web Intelligence offers two options to connect to SAP HANA. You can connect to SAP HANA by using a Web Intelligence MicroCube or you can connect online. If you decide to connect to SAP HANA via an online connection, any calculations will happen directly in SAP HANA instead of the Web Intelligence MicroCube.

In addition to direct connectivity, Web Intelligence is also able to leverage the Universe layer (see Figure 6.2).

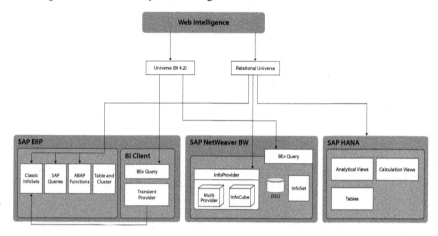

Figure 6.2 Web Intelligence and Universes

Using the Universe layer, Web Intelligence is able to offer additional connectivity options:

- A relational Universe can be used to connect to classic Infosets, ABAP queries, and ABAP functions in your SAP ERP system.

- A relational Universe can be used to set up connectivity to SAP HANA models and tables.

- A relational Universe can be established towards InfoProviders in SAP BW.

- Starting with SAP BusinessObjects BI 4.2, a Universe can also be created on top of BEx Queries.

CHAPTER 6

In this section we reviewed the different data-connectivity options available with Web Intelligence. Before we look into details on the supported metadata for SAP BW and SAP HANA we will discuss more details about the Universe layer as part of SAP BusinessObjects BI 4.2.

6.2 Semantic Layer and Data Connectivity

Before we discuss the details of data connectivity using Universes as part of your SAP BusinessObjects BI 4.2 system, I would like to clarify a couple of terms that are used in this book.

- The Semantic Layer is a technology that is part of the SAP BusinessObjects stack. It allows you to expose a given data source to your end users using more user-friendly business terms.

- Universes are artifacts created using the Information Design Tool as part of the SAP BusinessObjects BI platform. Universes represent a model using a data connection and business term. Universes can be leveraged by most of the BI client tools to expose information to end users.

- Dimensional Universes are one form of Universes focusing on multidimensional capabilities, such as hierarchies.

- Relational Universes are another form of Universes exposing the business terms in a flat view to end users.

- Multi-Source Universes allow users to combine multiple data sources into a single Universe and, in that way, expose these multiple sources in the form of a single logical view to end users.

- The Information Design Tool is the client tool that allows you to establish data connections towards the data sources and expose them in the form of business layers to your end users.

All of the above are elements and components of the overall Semantic Layer as part of the SAP BusinessObjects 4.2 environment. We will now continue and learn more about the different options to leverage the Semantic Layer in combination with your SAP system. SAP BusinessObjects 4.2 provides you with the option to expose a BEx Query directly to the BI client tools such as SAP BusinessObjects Analysis, edition for Microsoft Office (Analysis Office) or Web Intelligence.

Advantages of direct access to SAP BW:

- Allows re-use of existing BEx Queries.
- Offers shared connectivity across all BI client tools.
- Provides true hierarchical metadata and data.
- Allows for a single connection to point to multiple BEx Queries.
- Supports advanced BEx Query elements, such as restricted and calculated key figures, formulas, and custom structures.

Disadvantages of direct access to SAP BW:

- Does not allow for customizations of the metadata.
- Does not allow for the creation of custom objects.
- Does not allow for the creation of Universe-based parameters. (All parameters need to be either based on variables in the BEx Query or need to be created as parameters on a report level.)

In addition to the direct access to the SAP BW system, the Semantic Layer also provides you with the option to create a relational or Multi-Source Universe on top of SAP BW. Starting with SAP BusinessObjects 4.2 you also have the option to create a Universe on top of BEx Queries. The relational interface on top of SAP BW provides you with the option to create a Universe on top of the SAP BW system and also to combine data coming from SAP BW with other data sources using a Multi-Source Universe.

Advantages of relational access to SAP BW:

- Offers direct access to InfoProviders or the BEx Query level.
- Allows you to combine multiple data sources into a single logical view.
- Allows for customization of the metadata.
- Allows for creation of custom objects.

Disadvantages of relational access to SAP BW:

- Access to elements such as hierarchies, restricted and calculated key figures, BEx Query variables, and custom structures is only supported with the Universe on top of BEx Queries (available only starting with release 4.2).

- No support for Universes for Analysis Office.
- Limited support for Universes with Design Studio.

In the next section we will review the level of support for the SAP BW elements and compare the level of support for the three available data-connectivity options.

6.3 Supported and Unsupported SAP BW Elements

In this section we review the level of support for your existing metadata inside the SAP BW system for the three data-connectivity options for Web Intelligence as part of SAP BusinessObjects 4.2. We will compare direct access with the option to create a relational Universe and the option to set up a Universe on top of a BEx Query. Table 6.1 shows which of the objects are supported for these three options.

SAP BW Metadata	Direct Access Using BI Consumer Services (BICS)	Relational Universe	Universe Based on BEx Query (BI 4.2)
Direct access to InfoCubes and MultiProviders	Yes	Yes	Yes
Access to BEx Queries	Yes	No	Yes
Characteristic Values			
Key	Yes	Yes	Yes
Short description	Yes	Yes	Yes
Medium and long descriptions	Yes	Yes	Yes
BEx Query Features			
Support for hierarchies	Yes	No	Yes
Support for free characteristics	Yes	Yes	Yes
Support for calculated and restricted key figures	Yes	No	Yes

Table 6.1 Supported and unsupported BEx Query features for Web Intelligence *(continues)*

231

SAP BW Metadata	Direct Access Using BI Consumer Services (BICS)	Relational Universe	Universe Based on BEx Query (BI 4.2)
Support for currencies and units	Yes	No	Yes
Support for custom structures	Yes	No	Yes
Support for formulas and selections	Yes	No	Yes
Support for filters	Yes	No	Yes
Support for display and navigational attributes	Yes	No	Yes
Support for conditions in rows	No	No	No
Support for conditions in columns	No	No	No
Support for conditions for fixed characteristics	No	No	No
Support for exceptions	No	No	No
Compounded characteristics	Yes	Yes	Yes
Constant selection	Yes	No	Yes
Default values in BEx Queries	No	No	No
Number scaling factor	Yes	No	Yes
Number of decimals	No	No	No
Calculate rows as (local calculation)	No	No	No
Sorting	No	No	No
Hide/Unhide	Yes	No	Yes
Display as hierarchy	No	No	No
Reverse sign	Yes	No	Yes
Support for reading master data	Yes	No	Yes
Data Types			
Support for CHAR (characteristics)	Yes	Yes	Yes
Support for NUMC (characteristics)	Yes	Yes	Yes
Support for DATS (characteristics)	Limited	Yes	Limited
Support for TIMS (characteristics)	Limited	Yes	Limited
Support for numeric key figures such as Amount and Quantity	Yes	Yes	Yes
Support for Date (key figures)	Yes	Yes	Yes
Support for Time (key figures)	Limited	Yes	Limited

Table 6.1 Supported and unsupported BEx Query features for Web Intelligence (continues)

SAP BW Metadata	Direct Access Using BI Consumer Services (BICS)	Relational Universe	Universe Based on BEx Query (BI 4.2)
BEx Variables—Processing Type			
User input	Yes	No	Yes
Authorization	Yes	No	Yes
Replacement path	Yes	No	Yes
SAP exit/custom exit	Yes	No	Yes
Precalculated value set	Yes	No	Yes
General Features for Variables			
Support for optional and mandatory variables	Yes	No	Yes
Support for key date dependencies	Yes	No	Yes
Support for default values	Yes	No	Yes
Support for variable variants	No	No	No
Support for personalized values	No	No	No
BEx Variables—Variable Type			
Single value	Yes	No	Yes
Multi-single value	Yes	No	Yes
Interval value	Yes	No	Yes
Selection option	Limited	No	Limited
Hierarchy variable	Yes	No	Yes
Hierarchy node variable	Yes	No	Yes
Hierarchy version variable	Yes	No	Yes
Text variable	Yes	No	Yes
EXIT variable	Yes	No	Yes
Single key date variable	Yes	No	Yes
Multiple key dates	Yes	No	Yes
Formula variable	Yes	No	Yes

Table 6.1 Supported and unsupported BEx Query features for Web Intelligence (continued)

CHAPTER 6

In Table 6.2 you can see how the direct-access method using the BICS option and the Universe on top of the BEx Query uses elements from the BEx Query and how the objects are mapped to the Universe layer and direct access used by Web Intelligence.

BEx Query Element	Web Intelligence
Characteristic	For each characteristic you'll receive a field representing the key value and a field for the description, including short, medium, and long descriptions.
Hierarchy	Each available hierarchy is shown as an external hierarchy in Web Intelligence. At the point of defining the Web Intelligence query, the user needs to select the hierarchy levels or the hierarchy nodes that should be available in the report.
Key figure	Each key figure can have up to four elements: numeric value, unit, scaling factor, and formatted value. The formatted value is based on the user preferences configured in the SAP system.
Calculated/Restricted key figure	Each calculated and restricted key figure is treated like a key figure. The user does not have access to the underlying definition in Web Intelligence.
Filter	Filters are applied to the underlying query, but are not visible in Web Intelligence.
Display attribute	Display attributes become standard fields in the query panel and are grouped as subordinates of the linked characteristic.
Navigational attribute	Navigational attributes are treated the same way as characteristics.
Variable	Each variable with the property Ready for Input results in a parameter field in Web Intelligence.
Custom structure	A custom structure is available as an element in the query panel and each structure element can be selected or de-selected for the report.

Table 6.2 BEx Query metadata mapping for Web Intelligence

As you can see in Table 6.1 and Table 6.2, Web Intelligence is still lacking some key features when it comes to the connectivity with SAP BW, such as support for conditions, support for variable variants, full support for date- and time-based dimensions, and support for local calculations.

You will notice that the support of BEx Query elements in the relational Universe is even far less compared to direct connectivity. That brings up the question of why you would use the relational Universe on top of an InfoProvider instead of direct connectivity to the BEx Query layer. The relational Universe does provide you with one advantage, and that is the ability to combine several data sources with your data from SAP BW in a single Universe, and in that way provide your user with a single logical view on top of several data sources.

After reviewing the support for SAP BW elements, we will review the support for SAP HANA elements in the next section.

6.4 Supported and Unsupported SAP HANA Elements

In this section we review the level of support for your existing metadata inside SAP HANA for Web Intelligence. Table 6.3 shows which of the objects are supported when using Web Intelligence as the BI client tool.

SAP HANA Metadata	Direct Access to SAP HANA
Analytical Model	Yes
Calculation Model	Yes
Attribute Model	Yes
Dimension Key	Yes
Dimension description (Label column)	Yes
Unit/Currency	No
Calculated column	Yes
Restricted column	Yes
Variable	Yes
Input Parameter	Yes
Level-Based Hierarchy	Yes
Parent-Child Hierarchy	No

Table 6.3 Supported elements from SAP HANA

After reviewing the support for your SAP HANA-based metadata, we will learn how to set up a relational Universe for an SAP InfoProvider and how to create a Universe based on a BEx Query.

235

CHAPTER 6

6.5 Setting Up Universes

In addition to the option to leverage a direct connection towards the BEx Query using BICS, you can also establish a relational Universe on top of your InfoProviders in the SAP BW system, and you can set up a Universe based on a BEx Query. In the following steps, we show you how to create a Universe using a relational connection on top of an InfoProvider from the SAP NetWeaver Demo content. We will use the new option to create a Universe directly on top of the BEx Query.

Follow these steps to set up the Universe:

1. Start the Information Design Tool.

2. Select the menu File > New > Project to create a new project for your universe.

3. Enter a name for the new project and click Finish.

4. In the Repository Resources window, select the Insert Session menu to establish a session to your SAP BusinessObjects BI platform (see Figure 6.3).

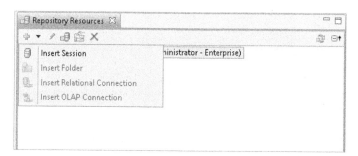

Figure 6.3 Repository resources

5. Enter a name for the new project and click Finish.

6. Log on to your SAP BusinessObjects BI platform.

7. Click OK.

8. Open the context menu of your established server connection in the Connections area (see Figure 6.4).

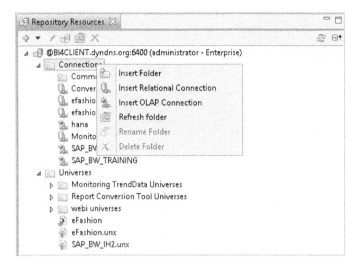

Figure 6.4 Insert the connection

9. Select the Insert Relational Connection menu item.

10. Enter a name for the connection.

11. Click Next.

12. Select the SAP BW connection type (see Figure 6.5).

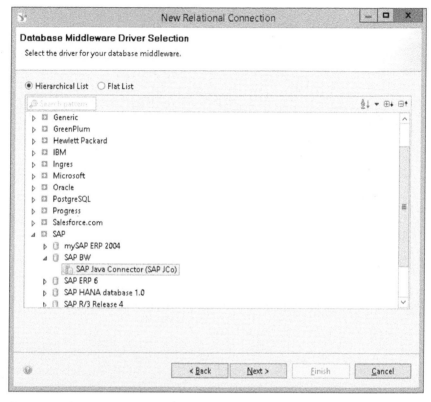

Figure 6.5 SAP BW connection

13. Click Next (see Figure 6.6).

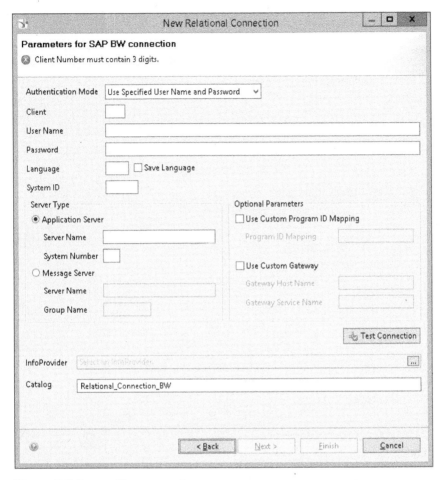

Figure 6.6 Connection parameters

14. Enter the necessary details of your SAP BW system:

- Client number
- User Name and Password
- Language
- System ID
- Application Server and System Number, or Message Server and logon Group Name

CHAPTER 6

Authentication Mode

You can set the authentication mode to Use Single Sign-On, but this requires your SAP BusinessObjects BI platform to be configured with SAP Authentication.

15. You can use the Save Language option to save your settings as configured in the relational connection. If you leave the check box unselected, the user can influence the language by setting the user preferences in the BI launch pad.

16. Use the ⌷ icon next to the InfoProvider field to receive a list of possible InfoProviders (see Figure 6.7).

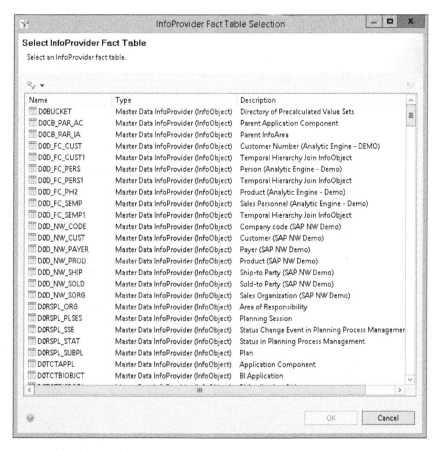

Figure 6.7 List of InfoProviders

17. You can use the filter as part of the screen to limit the list of InfoProviders based on the type of InfoProvider (see Figure 6.8).

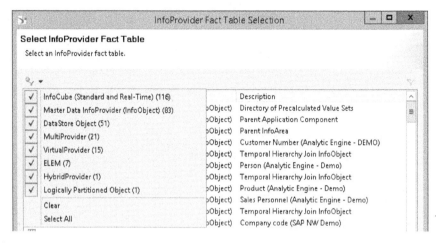

Figure 6.8 Filter option

18. In our example, we use the MultiProvider 0D_NW_M01 from the SAP NetWeaver Demo model.

19. Click OK.

20. Click Finish.

21. You are asked if you would like to create a shortcut for your connection. Click Yes.

22. Click Close.

SAP BW Star Schema

The following is only basic information about the tables shown in the list of available tables. We recommend that you consider the documentation on SAP BW for further details:

- The fact table of the InfoProvider is shown with the prefix I.
- Master data tables are shown with the prefix D.
- Text tables are shown with the prefix T.

23. Select your local project.

24. Select the menu File > New > Data Foundation.

25. Enter a name for the data foundation.

26. Click Next.

27. Select the Multi-Source Enabled option. (The connection towards SAP BW is not available when using the single-source option.)

28. Click Next. You are asked to log on to your SAP BusinessObjects BI platform.

29. Enter your credentials.

30. Click Next (see Figure 6.9).

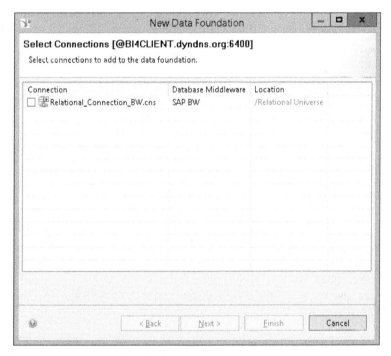

Figure 6.9 Connection shortcut

31. Select the shortcut that was created for the connection established previously.

32. Click Next.

33. Click Advanced (see Figure 6.10).

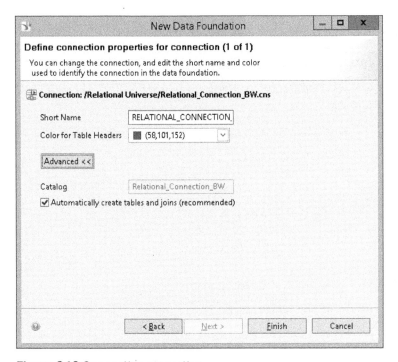

Figure 6.10 Connection properties

34. Ensure that the Automatically create tables and joins (recommended) option is activated.

35. Click Finish.

You are presented with a default-generated star schema for the selected InfoProvider.

36. Select your local project.

37. Select the menu File > New > Business Layer.

38. Select the Relational Data Source entry.

39. Click Next.

40. Enter a name for the business layer.

41. Click Next.

42. Select the previously generated relational Data Foundation (see Figure 6.11).

Figure 6.11 Select the data foundation

43. Ensure that the option Automatically create classes and objects for SAP NetWeaver BW connections (recommended) is activated.

44. Click Finish (see Figure 6.12).

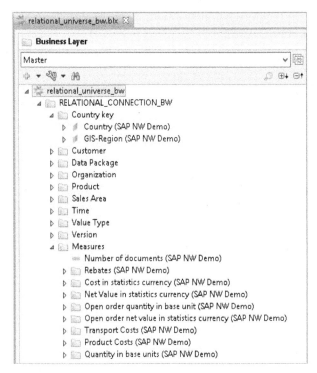

Figure 6.12 Generated business layer

You are presented with a list of classes, dimensions, and measures that have been generated based on the information retrieved from SAP BW.

45. Right-click the newly generated business layer entry as part of your local project (see Figure 6.13).

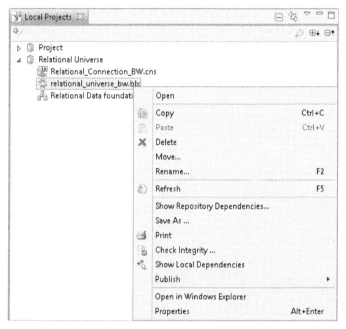

Figure 6.13 Context Menu

46. Select the menu Publish > To a Repository.

47. Select the integrity checks you would like to perform.

48. Click Next.

49. Select a folder for the Universe.

50. Click Finish.

51. Click Close.

You have created a Universe based on a relational connection on top of a MultiProvider in SAP BW. You can use Web Intelligence now on top of this Universe.

With the release of SAP BusinessObjects BI 4.2, you can now also create a Universe directly on top of a BEx Query and provide your users with a common layer on top of your existing SAP BW metadata. In the next set of steps we will use this new functionality to set up a Universe based on a BEx Query.

Follow these steps to set up a Universe based on a BEx Query:

1. Start the Information Design Tool.

2. Select the menu File > New > Project to create a new project for your Universe.

3. Enter a name for the new project and click Finish.

4. In the Repository Resources window select the Insert Session menu to establish a session to your SAP BusinessObjects BI platform.

5. Open the context menu of your established server connection in the Connections area.

6. Select the Insert OLAP Connection menu item.

7. Enter a name for the connection.

8. Click Next (see Figure 6.14).

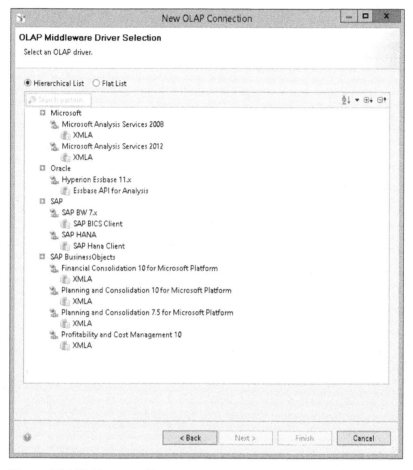

Figure 6.14 OLAP connections

9. Select the SAP BW 7.x connection type.

10. Click Next (see Figure 6.15).

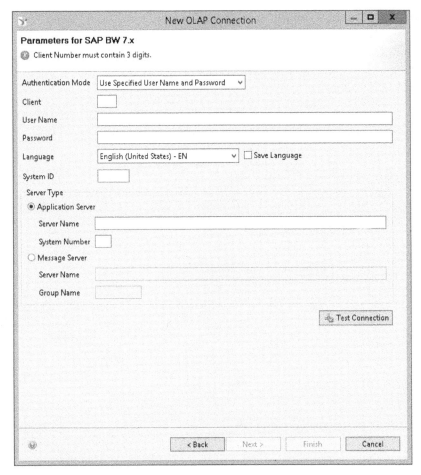

Figure 6.15 Connection parameters

11. Enter the necessary details of your SAP BW system:

- Client number
- User Name and Password
- Language
- System ID
- Application Server and System Number, or Message Server and logon Group Name

12. You can use the Save Language option to save your settings as configured in the relational connection. If you leave the check box blank, the user can influence the language by setting the user preferences in the BI launch pad.

13. Click Test Connection.

14. If the test fails, ensure that the connection details are correct.

15. Click OK to close the Test Connection message.

16. Click Next (see Figure 6.16).

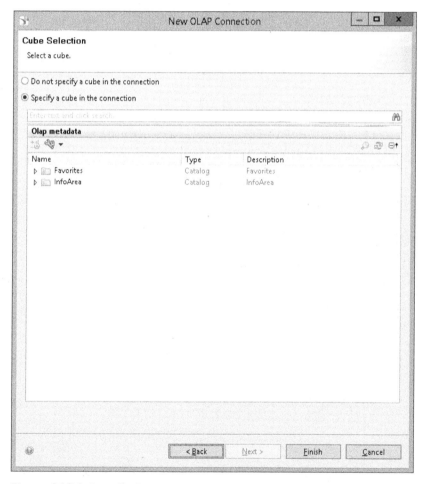

Figure 6.16 Cube selection

17. Select the option Do not specify a cube in the connection.

18. Click Finish.

19. Select the BW OLAP Connection in the list of connections and open the context menu with a right-click (see Figure 6.17).

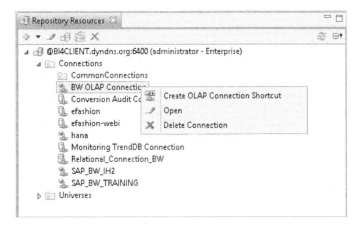

Figure 6.17 Repository Resources

20. Select the menu Create OLAP Connection Shortcut.

21. Select your local project in which to place the shortcut.

22. Click OK.

23. Select the newly created connection shortcut in your project and open the context menu with a right-click (see Figure 6.18).

Figure 6.18 Context Menu

24. Select the menu New Business Layer.

25. Enter a name for the New Business Layer.

26. Click Next (see Figure 6.19).

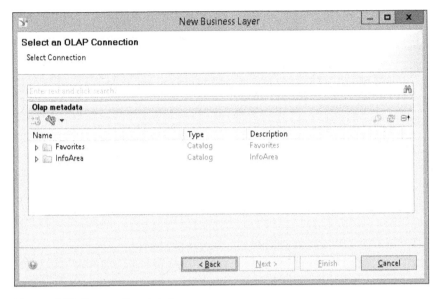

Figure 6.19 Connection details

27. You can now choose the BEx Query that you would like to use for the new Universe. You can use the search option or you can navigate to the InfoArea in your SAP BW system.

28. Select the BEx Query and click Next (see Figure 6.20).

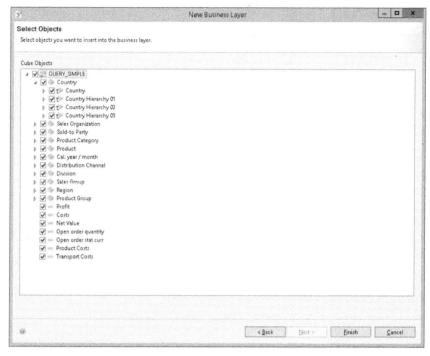

Figure 6.20 Object selection

29. You can now choose the elements from the available objects based on the selected BEx Query that you would like to include in the new Universe.

30. Select the objects and click Finish (see Figure 6.21).

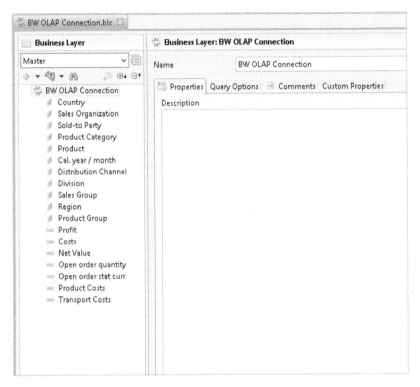

Figure 6.21 New Business Layer

31. You are presented with the new business layer based on the selected objects, and you can now publish your new Universe to the SAP BusinessObjects BI platform.

32. Right-click the newly generated business layer entry as part of your local project.

33. Select the menu Publish > To a Repository.

34. Select the integrity checks you would like to perform.

35. Click Next.

36. Select a folder for the Universe.

37. Click Finish.

38. Click Close.

Please note that you did not have to create a data foundation layer but instead you were able to create a business layer directly based on the BEx Query and you can now use this newly created Universe with Web Intelligence.

After creating the first set of Universes, we will create our first Web Intelligence report in the next section.

6.6 Creating Your First Web Intelligence Report

In the following steps you will learn how to use some of the basic functionality of Web Intelligence. You will leverage the connection created previously in the Central Management Console (CMC) of your SAP BusinessObjects BI platform system.

For our example steps we are using a BEx Query based on the SAP BW Demo model (http://scn.sap.com/docs/DOC-8941) with the following details:

Rows:

- Country (0D_NW_CNTRY)
- Region (0D_NW_REGIO)

Columns:

- Net Value (0D_NW_NETV)
- Costs (0D_NW_COSTV)

Free Characteristics:

- Product Group (0D_NW_PROD__0D_NW_PRDGP)
- Product Category (0D_NW_PROD__0D_NW_PRDCT)
- Cal. Year/Month (0CALMONTH)
- Calendar Year (0CALYEAR)

To create your first Web Intelligence report, follow these steps:

1. Log on to the SAP BusinessObjects BI launch pad.

2. Log on with your user credentials.

3. Navigate to the menu Applications (see Figure 6.22).

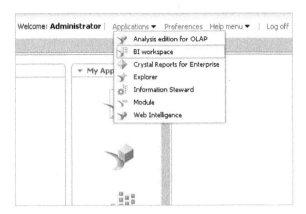

Figure 6.22 Applications

4. Select Web Intelligence.

5. Using the toolbar, start the process to create a new report (see Figure 6.23).

Figure 6.23 Web Intelligence toolbar

6. Select the option BEx as the data source (see Figure 6.24).

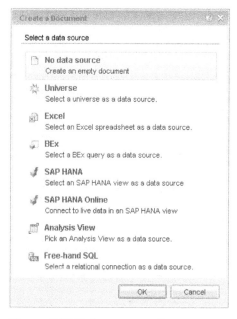

Figure 6.24 Data source options

7. Click OK (see Figure 6.25).

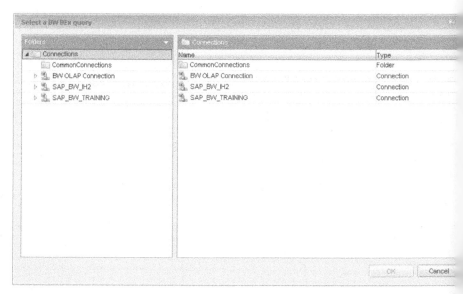

Figure 6.25 List of connections

8. In the list of connections, select the connection you created previously.

9. Double-click the connection.

10. Open the list of InfoAreas with a double-click.

11. Open the InfoArea for your InfoProvider.

12. Select the InfoProvider for your BEx Query with a double-click.

13. You will be presented with a list of BEx Queries.

14. Select the desired BEx Query.

15. Click OK (see Figure 6.26).

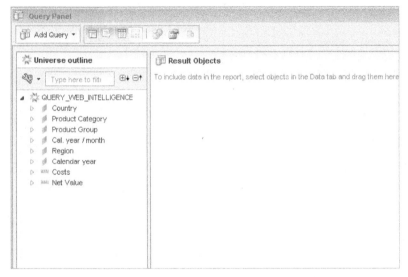

Figure 6.26 Web Intelligence Query Panel

16. Add the following dimensions and measures to the Result Objects of the Query Panel:

- Country
- Region
- Product Group
- Product Category
- Cal. Year/Month

- Net Value
- Costs

17. Open the Query Properties by clicking the highlighted icon (see Figure 6.27).

Figure 6.27 Query Panel Properties

18. Ensure the option Enable query stripping is activated (see Figure 6.28).

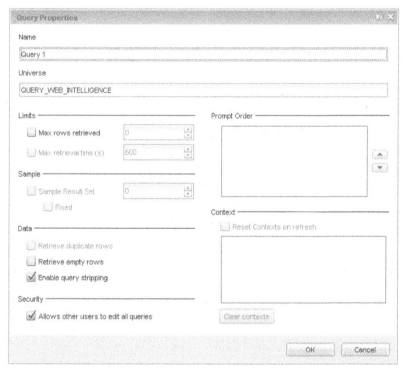

Figure 6.28 Query Properties

Query Stripping

Query stripping allows Web Intelligence to have a list of available characteristics, but not to ask for the data up front, similar to Free Characteristics in a BEx Query.

19. Click OK.

20. Click Run Query.

21. You are presented with a default layout for your report (see Figure 6.29).

Report 1

Country	Region	Product Group	Product Category	Cal. year / month	Net Value	Costs
United Arab Emirates	Abu Dhabi Emirate	MP3 & Headphones	Headphones	01.2012	375,967	333,854.11
United Arab Emirates	Abu Dhabi Emirate	MP3 & Headphones	Headphones	02.2012	405,260	359,501.49
United Arab Emirates	Abu Dhabi Emirate	MP3 & Headphones	Headphones	03.2012	442,110	393,661.38
United Arab Emirates	Abu Dhabi Emirate	MP3 & Headphones	Headphones	04.2012	485,762	431,772.8
United Arab Emirates	Abu Dhabi Emirate	MP3 & Headphones	Headphones	05.2012	480,551	428,602.67
United Arab Emirates	Abu Dhabi Emirate	MP3 & Headphones	Headphones	06.2012	383,279	340,531.49
United Arab Emirates	Abu Dhabi Emirate	MP3 & Headphones	Headphones	07.2012	448,056	398,244.72
United Arab Emirates	Abu Dhabi Emirate	MP3 & Headphones	Headphones	08.2012	389,981	346,534.61
United Arab Emirates	Abu Dhabi Emirate	MP3 & Headphones	Headphones	09.2012	451,477	401,464.03
United Arab Emirates	Abu Dhabi Emirate	MP3 & Headphones	Headphones	10.2012	1,009,880	898,675.16
United Arab Emirates	Abu Dhabi Emirate	MP3 & Headphones	Headphones	11.2012	725,250	646,997.34
United Arab Emirates	Abu Dhabi Emirate	MP3 & Headphones	Headphones	12.2012	788,356	701,225.08
United Arab Emirates	Abu Dhabi Emirate	MP3 & Headphones	MP3 Player	01.2012	1,919,510	1,427,193.17
United Arab Emirates	Abu Dhabi Emirate	MP3 & Headphones	MP3 Player	02.2012	2,176,240	1,589,183.01
United Arab Emirates	Abu Dhabi Emirate	MP3 & Headphones	MP3 Player	03.2012	2,121,435	1,601,490.81
United Arab Emirates	Abu Dhabi Emirate	MP3 & Headphones	MP3 Player	04.2012	1,854,361	1,366,685.23

Figure 6.29 Web Intelligence report

22. On the left side, navigate to the tab Properties and click Document (see Figure 6.30).

Figure 6.30 Document properties

23. The document properties are shown. Ensure that the option Enable query stripping is activated.

24. Click OK.

25. Select the column for Country by clicking the first country entry (not the column header).

26. Navigate to the tab Analysis.

27. Select the option Sort as part of the tab Display.

28. Select the option Advanced... (see Figure 6.31).

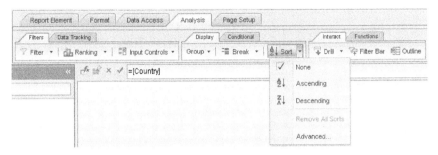

Figure 6.31 Sorting

29. Click Add.

30. Select Country.

31. Click OK.

32. Set the sort order to Descending.

33. Click Add

34. Select Region.

35. Click OK.

36. Set the sort order to Descending.

37. Click OK.

38. Select the first entry in the column Country (not the column header).

39. Navigate to the tab Analysis.

40. Select the tab Display.

41. Select the menu Break > Add Break (see Figure 6.32).

Figure 6.32 Add Break

42. Select the first entry in the column Region (not the column header).

43. Select the menu Break > Add Break.

44. Select the first entry in the column Cal. Year/Month (not the column header).

45. Right-click and open the context menu (see Figure 6.33).

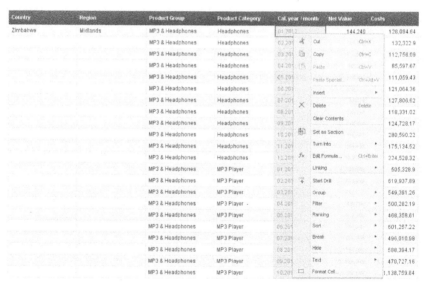

Figure 6.33 Set as Section menu item

46. Select the menu item Set as Section (see Figure 6.34).

01.2012

Country	Region	Product Group	Product Category	Net Value	Costs
Zimbabwe	Midlands	MP3 & Headphones	Headphones	144,240	128,084.64
		MP3 & Headphones	MP3 Player	668,030	505,328.9
		Cameras	Digital Compact Camer:	1,001,075	639,764.44
		Cameras	Digital SLR	2,172,353	1,688,282.64
		Laptops	Laptop	1,738,912	996,002.37
		Laptops	Mac Book	2,491,426	1,937,117.94
		Cell Phones	Android Phones	579,415	378,821.78
		Cell Phones	Blackberry OS	323,817	199,752.93
		Cell Phones	Apple iOS	391,951	287,358.88
		Cell Phones	Windows Phone	124,786	80,875.6
		Cell Phone Accessories	Cell Phone Cases	71,635	54,663.76
		Cell Phone Accessories	Handsfree Sets	166,471	98,641.8
		TV	LCD TV	2,809,309	2,000,460
		TV	Plasma TV	4,087,366	3,058,579.79
	Midlands				

Figure 6.34 Web Intelligence report

47. Select the first entry in the column Net Value (not the column header).

48. Navigate to the tab Analysis.

49. Select the tab Functions (see Figure 6.35).

Figure 6.35 Functions

50. Select the option Sum.

51. Repeat the steps for the column Costs to create the subtotals also for Costs.

52. Navigate to the tab Report Element.

53. Select the tab Chart (see Figure 6.36).

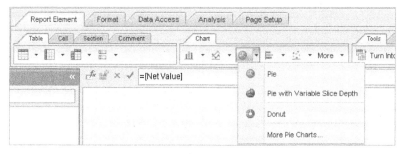

Figure 6.36 Charts

54. Select the Pie chart option.

55. Use a drag-and-drop navigation and place the pie chart on the right side of your table.

56. Now drag and drop the dimension Product Group from the list of available objects (left side) to the chart.

57. Drag and drop the measure Net Value from the list of available objects (left side) to the chart (see Figure 6.37).

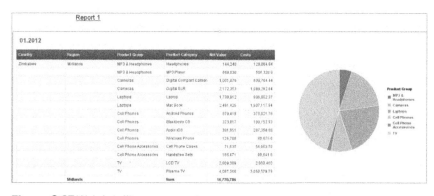

Figure 6.37 Web Intelligence report

58. Select the chart.

59. Navigate to the tab Report Element.

60. Navigate to the tab Position (see Figure 6.38).

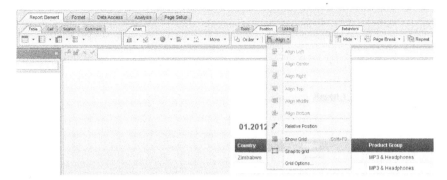

Figure 6.38 Menu position

61. Select the menu **Align > Relative Position.**

62. Ensure the Layout item in the Global area is selected (see Figure 6.39).

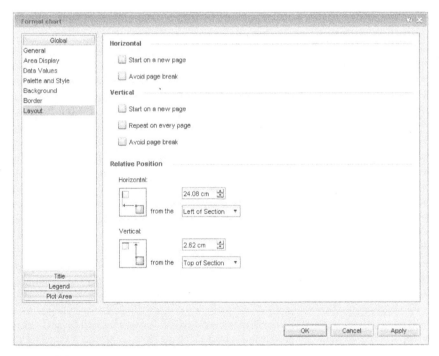

Figure 6.39 Format chart

63. Set the configuration as follows:

    - Horizontal
        - 1 cm / in from the Right Edge from Block 1 (your table)
    - Vertical
        - 1 cm / in from the Top Edge from Calendar Year/Month (your section)

64. Click OK.

65. Resize the columns of your table and the chart will now move accordingly without overlapping with the table.

66. Select the first entry in the column Country.

67. Navigate to the tab Analysis.

68. Navigate to the tab Interact (see Figure 6.40).

Figure 6.40 Interact tab

69. Select the option Outline.

70. You can now use the options to open and close the breaks from the table. You have the option on the top that will work for the individual items (for example, Country Zimbabwe), and then you have the items at the bottom that will work for all the areas (see Figure 6.41).

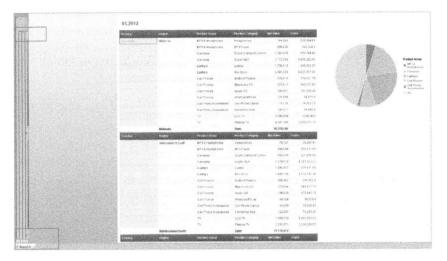

Figure 6.41 Outline interaction

71. Navigate to the tab Analysis.

72. Navigate to the tab Filters.

73. Click the small arrow on the right side of the Input Controls button (see Figure 6.42).

Figure 6.42 Input Controls option

74. Select the menu Define Control... (see Figure 6.43).

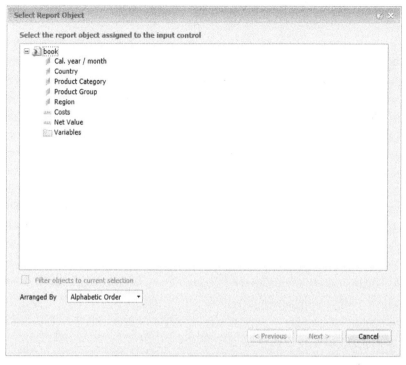

Figure 6.43 Select the report object

75. Select the entry Cal. Year/Month.

76. Click Next.

77. Select the option Radio buttons.

78. Ensure the option Allow selection of all values is activated (see Figure 6.44).

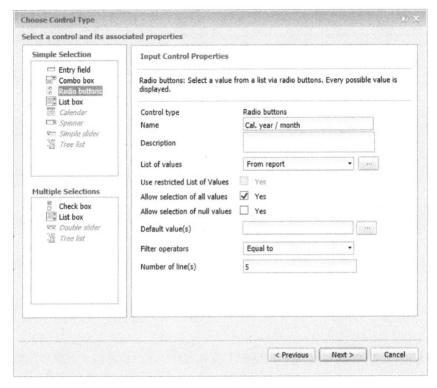

Figure 6.44 Select control

79. Click Next.

80. Select the section defined on Cal. Year/Month (see Figure 6.45).

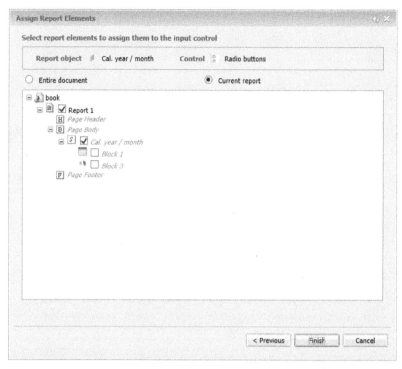

Figure 6.45 Select the report elements

81. Click Finish.

82. You can now use the Input Control on the left side and select the different months to filter the data in the report (see Figure 6.46).

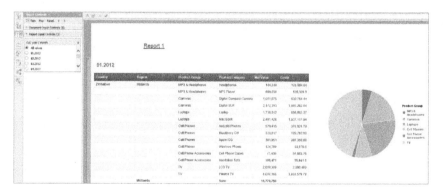

Figure 6.46 Input controls

83. Select the first value in the column Product Group.

84. Use a right-click and use the option Delete.

85. Select a value in the column Product Category.

86. Use a right-click and use the option Delete.

87. Refresh your report.

88. Take a look at the list of available objects on the left side (see Figure 6.47).

Figure 6.47 Available Objects

Because query stripping is enabled, the characteristic Product Category is now shown in bold as the data is not retrieved because it is not needed for the report. Product Group is not shown in bold as it is used in the chart.

89. Now drag the characteristic Product Category back to the table.

90. Because the data is not in the report, the report needs to be refreshed.

91. Refresh your report.

92. Navigate to the tab File (left side).

93. Select the menu Save (see Figure 6.48).

Figure 6.48 Menu file

94. Select a folder for your report.

95. Enter a name for the report and click Save.

You can now log on to the BI launch pad and view and share your report.

In this section you created your first Web Intelligence report using the direct-access method leveraging BICS connecting to a BEx Query.

6.7 Summary

In this chapter you learned how you can leverage Web Intelligence in combination with data from your SAP BW and SAP HANA systems. You also learned how the product supports the existing metadata from your SAP BW system and your SAP HANA system, and how you can set up Universes on top of your SAP system. In the next chapter we will review some of the upcoming roadmap changes.

Integration Roadmap Update

After reviewing the different integration options of the SAP BusinessObjects BI clients in combination with SAP BW and SAP HANA, we will take a look at some of the upcoming changes and enhancements to the SAP BusinessObjects BI portfolio.

Product Roadmap Disclaimer

The descriptions of future functionality in this chapter are the author's interpretation of the publicly available product integration roadmap. These items are subject to change at any time without any notice, and the author does not provide any warranty on these statements.

7.1 Overall Strategic Direction

In regards to the overall BI offerings from SAP, there are two main directions that SAP is moving towards: on premise and cloud. Figure 7.1 shows these two main directions and their components.

CHAPTER 7

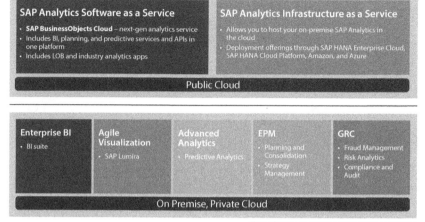

This graphic is based on publicly available information from SAP.

Figure 7.1 On premise versus the cloud

It is very important to acknowledge that SAP is proceeding with both directions; therefore, it is up to the customer to decide whether to move to a cloud-based environment or to continue with an on-premise landscape. As shown in Figure 7.1, the SAP BusinessObjects BI platform is the leading environment for on-premise customers and SAP BusinessObjects Cloud (former SAP Cloud for Analytics) is the leading environment for cloud-based scenarios.

Figure 7.2 shows the overall plan for simplifying the BI portfolio and which BI clients from the overall SAP BusinessObjects BI portfolio are part of SAP's strategic direction.

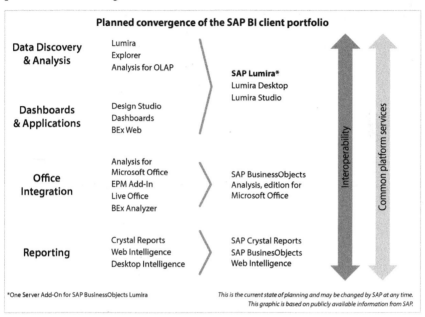

Figure 7.2 BI simplification

As shown in Figure 7.2, SAP BusinessObjects Analysis, edition for Microsoft Office (Analysis Office), SAP BusinessObjects Lumira, Design Studio, Crystal Reports, and Web Intelligence are the BI clients that are part of SAP's future plans. The products shown on the left side—for example, SAP BusinessObjects Explorer—are seen as legacy products, and companies should consider moving to the new BI clients as part of their own strategic direction.

In the next sections, we will review some of the more detailed roadmaps for some of the BI clients.

CHAPTER 7

7.2 Analysis Office

With Analysis Office as the leading product when it comes to the integration with the Microsoft Office environment, it is important to understand that SAP is planning to consolidate Analysis Office with the former Enterprise Performance Management (EPM) plug-in that is in use by customers for the SAP BusinessObjects Planning and Consolidation (BPC) product line. In addition, Analysis Office is already the replacement for the BEx Analyzer product, and by now most customers have moved their workbooks from BEx Analyzer to Analysis Office. Figure 7.3 shows the upgrade path options for those customers that are currently using the EPM plug-in.

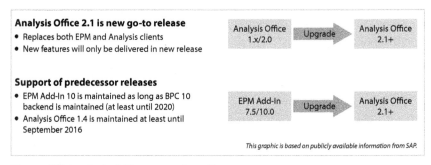

Figure 7.3 Analysis Office upgrade path

Figure 7.4 provides a recommendation on which client—Analysis Office or the EPM plug-in—to use, based on your existing landscape.

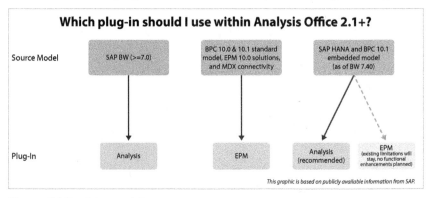

Figure 7.4 Decision matrix

With the latest release of Analysis Office (Release 2.3) SAP added functionality such as:

- Support for the Report-to-Report Interface as the receiver.

- Ability to create restricted measures directly in the workbook.

- Ability to rename dimensions and measures directly in the workbook.

- Support for BPC in Analysis Office.

Figure 7.5 shows the roadmap for Analysis Office combined with the EPM plug-in.

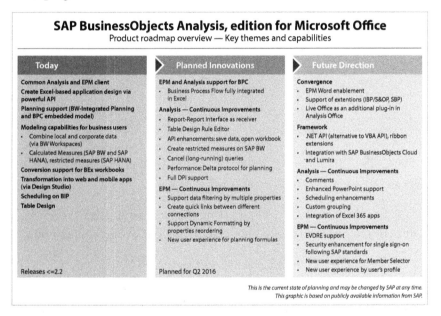

Figure 7.5 Analysis Office roadmap

As you can see in Figure 7.5, some very important topics are planned for the next release of Analysis Office (Future Direction):

- Integration of Live Office as an additional plug-in for Analysis Office.

- Integration of Analysis Office with SAP Lumira and SAP BusinessObjects Cloud.

- Integration with Microsoft Office 365 applications.

- Enhancements in the area of commenting and scheduling.

CHAPTER 7

In this section we reviewed the roadmap of Analysis Office; in the next section we will learn more about the futures of Design Studio and SAP Lumira.

7.3 Design Studio and SAP Lumira

In May 2016, SAP announced the combination of Design Studio and SAP Lumira into one product (see Figure 7.6).

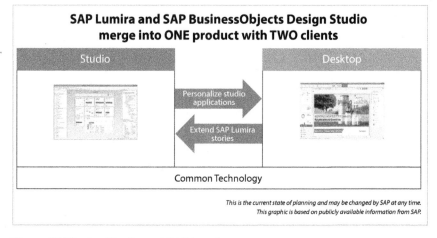

Figure 7.6 Design Studio and SAP Lumira

As shown in Figure 7.6, Design Studio and SAP Lumira are referred to as one product, but the important part is that it is actually one product with two BI clients. In other words, Design Studio and SAP Lumira will continue as two separate products with different target audiences, but SAP will brand these two products under one product family—SAP Lumira.

Figure 7.7 shows this convergence of products a little more clearly, by showing the two new product names: SAP Lumira Discovery and SAP Lumira Designer. Some highlights of the upcoming changes are:

- Ability to open SAP Lumira content in Design Studio, and open Design Studio in SAP Lumira.

- Consolidation of both server add-ons into a single add-on for the SAP BusinessObjects BI platform.

- Stronger integration of SAP Lumira with SAP BW and SAP HANA.

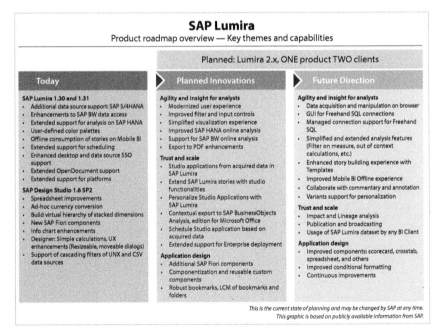

Convergence of Lumira and Design Studio
Lumira 2.0

SAP BusinessObjects Lumira
SAP BusinessObjects Explorer

SAP BusinessObjects Design Studio
SAP BEx Web Application Designer
SAP BusinessObjects Analysis for OLAP
SAP BusinessObjects Dashboards

Lumira 2.x
SAP Lumira Discovery
(formerly Lumira Desktop)
SAP Lumira Designer
(formerly Design Studio)
Single server-side add-on

Data Discovery & Applications

Benefits
- One product, two desktop experiences for the business user and the IT persona
- IT and power users can promote business user visualizations to enterprise dashboards
- Business users can personalize enterprise dashboards
- Changing requirements doesn't need a tool switch
- Singer server-side add-on reduces IT administration cost
- Improved ease of use and adoption with single canvas for both visualization and storyboard authoring
- Align with S/4 UI standard

This graphic is based on publicly available information from SAP.

Figure 7.7 SAP Lumira and Design Studio convergence

Figure 7.8 shows some of the additional highlights of the planned release.

SAP Lumira
Product roadmap overview — Key themes and capabilities

Planned: Lumira 2.x, ONE product TWO clients

Today	Planned Innovations	Future Direction
SAP Lumira 1.30 and 1.31 • Additional data source support: SAP S/4HANA • Enhancements to SAP BW data access • Extended support for analysis on SAP HANA • User-defined color palettes • Offline consumption of stories on Mobile BI • Extended support for scheduling • Enhanced desktop and data source SSO support • Extended OpenDocument support • Extended support for platforms **SAP Design Studio 1.6 SP2** • Spreadsheet improvements • Ad-hoc currency conversion • Build virtual hierarchy of stacked dimensions • New SAP Fiori components • Info chart enhancements • Designer: Simple calculations, UX enhancements (Resizeable, moveable dialogs) • Support of cascading filters of UNX and CSV data sources	**Agility and insight for analysts** • Modernized user experience • Improved filter and input controls • Simplified visualization experience • Improved SAP HANA online analysis • Support for SAP BW online analysis • Export to PDF enhancements **Trust and scale** • Studio applications from acquired data in SAP Lumira • Extend SAP Lumira stories with studio functionalities • Personalize Studio Applications with SAP Lumira • Contextual export to SAP BusinessObjects Analysis, edition for Microsoft Office • Schedule Studio application based on acquired data • Extended support for Enterprise deployment **Application design** • Additional SAP Fiori components • Componentization and reusable custom components • Robust bookmarks, LCM of bookmarks and folders	**Agility and insight for analysts** • Data acquisition and manipulation on browser • GUI for Freehand SQL connections • Managed connection support for Freehand SQL • Simplified and extended analysis features (Filter on measure, out of context calculations, etc.) • Enhanced story building experience with Templates • Improved Mobile BI Offline experience • Collaborate with commentary and annotation • Variants support for personalization **Trust and scale** • Impact and Lineage analysis • Publication and broadcasting • Usage of SAP Lumira dataset by any BI Client **Application design** • Improved components: scorecard, crosstab, spreadsheet, and others • Improved conditional formatting • Continuous improvements

This is the current state of planning and may be changed by SAP at any time.
This graphic is based on publicly available information from SAP.

Figure 7.8 SAP Lumira roadmap

These highlights include:

- Ability to schedule Design Studio applications.

- Ability to re-use and personalize components in Design Studio.

- Overall modernized user experience in SAP Lumira and the consolidation of the Visualize and Compose rooms into a single visualization canvas.

After reviewing the Design Studio and SAP Lumira roadmap, in the next section we will take a look at the SAP BusinessObjects BI platform roadmap.

7.4 SAP BusinessObjects BI Platform

After reviewing the upcoming highlights of some of the BI clients, let's now take a look at some of the key enhancements for the SAP BusinessObjects BI platform (see Figure 7.9).

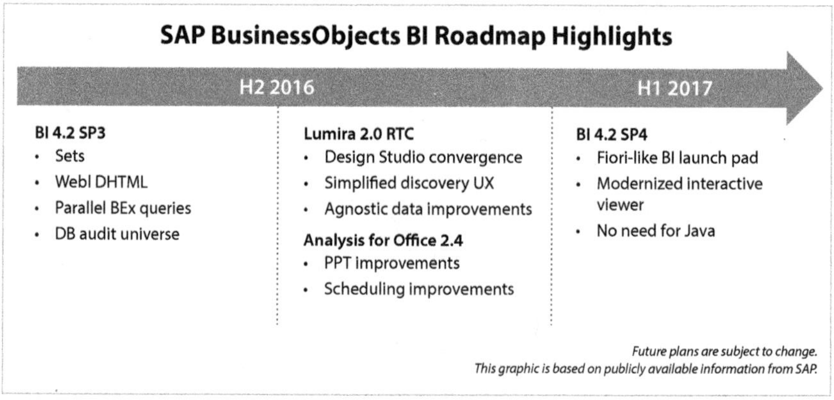

SAP BusinessObjects BI Roadmap Highlights

H2 2016		H1 2017
BI 4.2 SP3 • Sets • Webl DHTML • Parallel BEx queries • DB audit universe	**Lumira 2.0 RTC** • Design Studio convergence • Simplified discovery UX • Agnostic data improvements **Analysis for Office 2.4** • PPT improvements • Scheduling improvements	**BI 4.2 SP4** • Fiori-like BI launch pad • Modernized interactive viewer • No need for Java

Future plans are subject to change.
This graphic is based on publicly available information from SAP.

Figure 7.9 SAP BusinessObjects BI platform roadmap

Figure 7.9 shows the overall BI roadmap with the highlights for the SAP BusinessObjects BI 4.2 Service Pack 03 and Service Pack 04 releases. Some of the highlights for the upcoming Service Packs:

- Support for parallel execution of BEx Queries.
- Ability to add Sets to a Universe using the Information Design Tool.
- Ability to connect to the Central Management Service (CMS) database using an Open Database Connectivity (ODBC) Driver.
- Fiori-style launchpad.
- Improvements on the Microsoft SharePoint integration.

7.5 Summary

In this chapter, we reviewed some of the key highlights of the upcoming product releases to provide you with a brief outlook on some of the expected enhancements so that you can integrate those into your future plans for your SAP BusinessObjects BI landscape.

Conclusion

With this book, I have provided a straightforward, practical overview of the key aspects of the integration of SAP BusinessObjects BI with SAP software. We've looked at the SAP BusinessObjects BI platform and how to install and configure the 4.2 release of the platform and client components. We've covered SAP BusinessObjects Analysis, edition for Microsoft Office (Analysis Office), SAP BusinessObjects Design Studio, SAP BusinessObjects Lumira, and SAP BusinessObjects Web Intelligence in depth, and walked through how to create reports or dashboards with these products.

You've also gained a better understanding of the roadmap of the SAP BusinessObjects BI portfolio and how the upcoming changes will be able to enhance your SAP landscape.

Notes

insider BOOKS

Gain a thorough understanding of how to use
SAP technology in real-world scenarios.

Continue reading this book online as it's updated to build on the
core knowledge you've already gained. Follow along with the author,
step-by-step, as technology evolves.

Notes